AF787314

IntechOpen

Supporting open minds since 2005

Engineered Nanomaterials - Health and Safety
http://dx.doi.org/10.5772/intechopen.83105
Edited by Sorin Marius Avramescu, Kalsoom Akhtar, Irina Fierascu, Sher Bahadar Khan, Fayaz Ali and Abdullah M. Asiri

Contributors

Libor Kvítek, Ales Panacek, Robert Prucek, Jana Soukupova, Ioana Berindan-Neagoe, Huei Ruey Ong, Wan Mohd Eqhwan Iskandar, Md. Maksudur Rahman Khan, Tooba Saeed, Abdul Naeem, Tahira Mahmood, Nazish Huma Khan, Takalani Cele, Raul Alberto Morales Luckie, Edith Lara-Carrillo, Saraí C. Guadarrama-Reyes, María G. González-Pedroza, Ulises Velazquez-Enriquez, Víctor Toral-Rizo, Víctor Sanchez Mendieta, Rogelio Jose Scougall-Vilchis, Md. A. Mallick, Puja Kumari, Paritosh Patel, Suresh K Verma, Atamjit Singh, Kirandeep Kaur, Nicole Nelson, Kelly McGuire, David Busath, Jon Hogge, Aidan Hintze, Nathan Liddle, Jordan Pollock, Austin Brown, Steven Dee Walker, Johnny Lynch, Roger Harrison, Anca Onaciu, Ancuta Jurj, Cristian Moldovan, Stephen Facer

First published in London, United Kingdom, 2020 by IntechOpen
IntechOpen is the global imprint of INTECHOPEN LIMITED, registered in England and Wales, registration number: 11086078, 7th floor, 10 Lower Thames Street, London, EC3R 6AF, United Kingdom
Printed in Croatia

British Library Cataloguing-in-Publication Data
A catalogue record for this book is available from the British Library

Additional hard and PDF copies can be obtained from orders@intechopen.com

Engineered Nanomaterials - Health and Safety
Edited by Sorin Marius Avramescu, Kalsoom Akhtar, Irina Fierascu, Sher Bahadar Khan, Fayaz Ali and Abdullah M. Asiri
p. cm.
Print ISBN 978-1-83880-411-4
Online ISBN 978-1-83880-412-1
eBook (PDF) ISBN 978-1-83881-088-7

For EU product safety concerns:
IN TECH d.o.o., Prolaz Marije Krucifikse Kozulić 3, 51000 Rijeka, Croatia, info@intechopen.com or visit our website at intechopen.com.

Meet the editors

Dr. Sorin Marius Avramescu is a lecturer PhD at the University of Bucharest, Department of Organic Chemistry, Biochemistry, and Catalysis. He has been working in the environmental protection, multifunctional materials development, and analytical chemistry fields for 15 years and has a background of leading and managing large multidisciplinary, multi-partner research projects. He has been the author/co-author of more than 30 articles indexed in the Web of Science, Scopus, and other databases, 3 book chapters, and more than 60 participations in national and international scientific conferences. He has also filed 4 patents regarding materials development for ecofriendly processes.

Dr. Kalsoom Akhtar received her Ph.D from the Chemistry Department, Ewha Womans University, Seoul, Korea. Dr. K. Akhtar is an Assist. Professor in the Chemistry Department, King Abdulaziz University and is currently doing research in organic and nano-chemistry, which comprises photo-catalyst, organic synthesis, and metal oxide nanomaterials. She is the author of 2 books and 65 research papers.

Irina Fierascu is a scientific researcher (CS1) at INCDCP-ICECHIM Bucharest with competences in analytical chemistry (different modern analytical techniques), nanomaterials – obtaining and characterization, nanotechnology, materials science, archaeometry, superior use of natural resources (obtaining and using natural extracts), photosynthesized nanomaterials, environmental protection, etc. Her PhD is in Material Engineering, and she has followed postdoctoral programs in Industrial Biotechnology and Environmental Engineering. The habilitation thesis covers "Capitalization of autochthonous vegetal species: from biomedical applications to nanotechnology". She has been the author/co-author of several papers published in prestigious ISI journals (more than 80) in the field of natural extracts, nanomaterials, and photosynthesized metallic nanoparticles (especially gold and silver nanoparticles), as well as author/co-author of 12 books/chapter in books. She has experience in managing projects with topics of international relevance: application of nanomaterials in different fields: environment, agriculture or conservation/restoration of cultural heritage artifacts. She has won different awards, from government institutions or other prestigious institutions in the field.

Prof. Dr. Sher Bahadar Khan received his Ph.D from HEJ, Karachi University, Pakistan. After completion of his Ph.D, he started his post-doctoral career in nanochemistry and nanotechnology and continued to work as a post-doctoral research fellow to February 2010 at Yonsei University, South Korea. In March 2010, he joined the Center for Advanced Materials and Nano-engineering, Department of Chemistry, Najran University as an Assistant Professor and continued his work to 31 August 2011. He joined the Chemistry Department, King Abdulaziz University, Jeddah, Saudi Arabia as an Assistant Professor in September 2011. Currently he is Full Professor in the Chemistry Department, King Abdulaziz University and is doing research in nanochemistry and nanotechnology. He was honoured by receiving the top scientist award of KP Science & Technology in

2018. He was also honoured by the Deanship of Scientific Research awards at King Abdulaziz University for book, patent, and highly ranked scientific publication. He is the author of 320 research articles, twelve books, and six patents with almost 1000 ± 10 impact factor, 6665 citations, and 45 h-index.

Fayaz Ali was born in Mardan, Pakistan, and completed his B.Sc. and M.Sc. studies at Islamia College University Peshawar. He obtained his M.Phil. degree in Physical Chemistry from NCEPC, Peshawar, and then completed his PhD degree in 2018 from King Abdulaziz University, Saudi Arabia. He was awarded the Best Graduate Student Excellence Award of 2018 in PhD by the Deanship of Graduate Studies, KAU. He worked as Assistant Professor of Chemistry for one year at the Department of Chemistry, Abbottabad University of Science and Technology. Currently, he is working as a postdoctoral research fellow in the laboratory of Professor Yi Zhun Zhu at Macau University of Science and Technology, Macau. His research focus is on the design, synthesis, and applications of nanoscale materials.

Prof. Abdullah Mohamed Asiri received his PhD from the University of Wales, College of Cardiff, UK in 1995. He has been the Head of the Chemistry Department at King Abdulaziz University since October 2009 and he is the Founder and the Director of the Center of Excellence for Advanced Materials Research (CEAMR). He is a Professor of Organic Photochemistry. His research interests cover color chemistry, synthesis of novel photochromic, thermochromic systems, synthesis of novel colouring matters and dyeing of textiles, materials chemistry, nanochemistry, nanotechnology, polymers, and plastics. He is the Editor-in-Chief of King Abdulaziz University Journal of Science and is also a member of the Editorial Board of Pigments and Resin Technology (UK), Organic Chemistry in Sight (New Zealand), Recent Patents on Materials Science (USA). He is the Vice-President of the Saudi Chemical Society (Western Province Branch).

Contents

Preface

Nanotechnologies are extremely diverse, bringing about new opportunities in human lives through countless applications. There are numerous ways to produce dispersed nanomaterials in different shapes, forms, and functionalization, and our perception is challenged in terms of apprehending the complexity and diversity of nanostructures. Materials reduced to small dimensions are considerably different from molecules in a solution, bulk, or crystalline form and this opens the door to emerging new materials with very unique properties. It is for this reason that nanoparticles have made an invasive presence in the agriculture, medicine, and electronics industries, contributing to the unprecedented development of these fields. Nanoparticles prepared by different techniques and chemical composition bring about the question of the delicate balance between their importance and cautions for using them without proper assessment. This book is intended to emphasize a new perspective of knowledge on the environmental and human health impact of engineered nanoparticles in general with a focus on Ag nanoparticles as the most studied and manufactured material in this field. This book shows that friends or foes, nanoparticles can be managed and developed in environmentally benign ways to keep their outstanding features that drive the scientist to create them in the first place. Editing this book was a meaningful experience and also beneficial to improve knowledge of the field. The authors are renowned specialists from different countries and their expertise allows us to fulfill the difficult task of presenting some insightful data from this vast field. Thus, this study can be considered an important reference for chemists, biochemists, physicians, and materials scientists working with and developing nanoparticle systems with a focus on the possible impact on human health.

It has been a good experience to edit this book and interact with the other authors in the related fields, who are specialized in their area. I strongly acknowledge all of my partner editors, especially Dr. Sher Bahadar Khan for all of his support and precious time.

Sorin Marius Avramescu
Faculty of Chemistry,
Department of Organic Chemistry, Biochemistry and Catalysis,
University of Bucharest,
Bucharest, Romania

Kalsoom Akhtar, Sher Bahadar Khan and Abdullah Asiri
King Abdulaziz University,
Kingdom of Saudi Arabia

Irina Fierascu
National Institute for Research and Development in Chemistry and Petrochemistry,
Bucharest, Romania

Fayaz Ali
School of Pharmacy,
Macau University of Science and Technology,
Macau, China

Section 1

Preparation and Characterisation of Nanoparticles

Chapter 1

Green Synthesis of Silver Nanoparticles Using *Heterotheca inuloides* and Its Antimicrobial Activity in Catgut Suture Threads

Saraí C. Guadarrama-Reyes, Raúl A. Morales-Luckie, Víctor Sánchez-Mendieta, María G. González-Pedroza, Edith Lara-Carrillo, Ulises Velazquez-Enriquez, Victor Toral-Rizo and Rogelio Scougall-Vilchis

Abstract

Silver nanoparticles were synthesized through a green method, using *Heterotheca inuloides* as a bioreducing agent. Moreover, catgut suture threads were decorated with those biogenic silver nanoparticles, and their antibacterial activity versus highly resistant pathogenic microorganisms was evaluated. The principles of green chemistry and nanotechnology allow us to obtain advanced materials, such as suture threads, which can reduce or avoid the prevalence of infectious processes in the medical field. Mexican medicinal plants, such as *H. inuloides*, represent an adequate alternative for biosynthesis; this plant species is known for its medicinal benefits and its antibacterial activity, and for that reason, it is being used in folk medicine.

Keywords: *Heterotheca inuloides*, green synthesis, silver nanoparticles, antimicrobial activity, catgut, suture

1. Introduction

Diverse green synthesis methods, involving the use of plant extracts as reducing agents, provide attractive approaches to synthesize AgNPs.

Mexican medicinal plants represent an adequate alternative for biosynthesis, such is the case of *Heterotheca inuloides*, a plant known for its medicinal benefits, as well as anti-inflammatory and analgesic properties. The plant, commonly named as Mexican arnica has been traditionally used for its antimicrobial activity, antifungal, cytotoxic and antioxidative properties, leading the World Health Organization (WHO) to recognize its use in medicine. This species has also been used to treat dental diseases and gastrointestinal disorders [1–6].

The wide use of *H. inuloides*, in medicine, can be attributed to its more than 140 components. Several constituents of the aqueous extract obtained from the dried flowers have been identified and characterized as antibacterial agents. Flavonoids, sesquiterpenoids, triterpenoids, and sterols are mainly present on its chemical composition [7–9].

Conventional approaches in nano-synthesis involve the use of highly toxic chemicals, resulting in side effects after administration [10, 11]. For this reason, it is of utmost importance for biomedical science to try to minimize any consequent risk to human health.

In modern surgery, attempts have been made to reduce the prevalence of infections related with surgical sutures, through the use of coated materials [12]. Nevertheless, the risk of surgical site infection is a constant challenge in wound closure. By using sutures with an antibacterial coating, the risk of infection is considerably reduced. The significant feature of silver is its broad antimicrobial spectrum associated with biomaterial-related infections [11, 13].

We present a total green synthetic method where *Heterotheca inuloides* is used for the first time to decorate catgut, a suture thread widely used in surgery. Its characterization, and antimicrobial activity against *Staphylococcus aureus* and *Escherichia coli*, is reported.

2. Experimental

2.1 Synthesis of AgNPs

The plant material was collected from surrounding fields of the State of Mexico, and it was cleaned using tap water followed by distilled water. *H. inuloides* leaves were dried and finely ground to a powder and then kept at room temperature for 24 h. About 1 gram of powder was poured in 100 mL of distilled water and boiled. The solution was filtered using filter paper. A 10 mM silver nitrate solution ($AgNO_3$, Sigma-Aldrich) was prepared to generate AgNPs; both solutions were mixed in a 1:2.5 ratio.

After 6 h, catgut (USP 3-0, Atramat®) was totally immersed in the solution for 1 h and then taken out and dried at room temperature.

2.2 Characterization of AgNPs

2.2.1 Spectrophotometry by UV: Vis

UV-Vis was performed every hour for the next 6 h. Spectral measurements were recorded on a Cary 5000 UV-Vis-NIR Scanning Spectrophotometer using a quartz cell, and the wavelength ranges from 300 to 600 nm.

2.2.2 Spectrophotometry by FTIR

The FTIR analysis was performed (Bruker, Model 27) to identify the main functional groups in the aqueous extract of *Heterotheca inuloides*.

2.2.3 Scanning electron microscopy (SEM)

Catgut samples were prepared for its analysis in a JSM-6510-LV microscope (JEOL Tokyo, Japan) at 20 kV of acceleration and using secondary electrons.

The samples were coated with a thin film of gold (c.a. 20 nm) using a Denton Vacuum DESK IV sputtering equipment.

2.2.4 Transmission electron microscopy (TEM)

Transmission electron microscope (TEM, JEOL JEM-2100, Tokyo, Japan) was used. To evaluate shape and size of silver nanoparticles, the solution was analyzed by placing a drop on a copper grid (300 mesh) coated with carbon film and let to dry at room temperature. A 200 kiloelectronvolt accelerating voltage was used in bright-field mode and high resolution.

2.3 Antibacterial activity

The antibacterial activity of AgNPs was determined by well diffusion method against the *Staphylococcus aureus* and *Escherichia coli* on the Mueller-Hinton agar plates.

Catgut suture threads were cut into pieces of approximately 10 mm of length and put on the Petri dishes. Each plate was prepared in triplicate. The plates were incubated at 37°C in a Felisa® incubator for 24–48 h.

After incubation, a clear zone appeared, and by measuring the halo of inhibition for both strains, the antibacterial effect was assessed.

3. Results

3.1 UV-Vis spectroscopy

AgNPs synthesized by *Heterotheca inuloides* produced polydisperse and stable nanoparticles as shown in **Figure 1**. The increase in the intensity of surface plasmonic resonance, at 451 nm, as a function of time, is observed. In addition, it is confirmed that at 6 h, the formation of the nanoparticles is finished.

By means of transmission electron microscopy (TEM), the size and shape of AgNPs are demonstrated; an average nanoparticle size of 16.0 nm and a standard

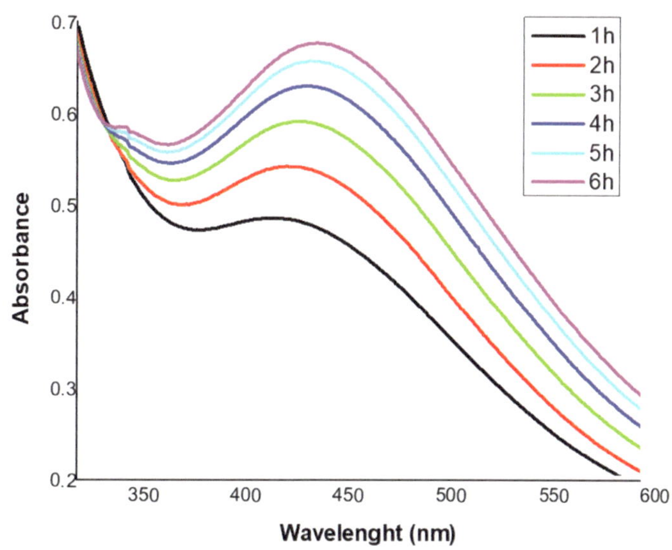

Figure 1.
UV-Vis spectra showing that the AgNP plasmon wavelength lies between 440 and 460 nm.

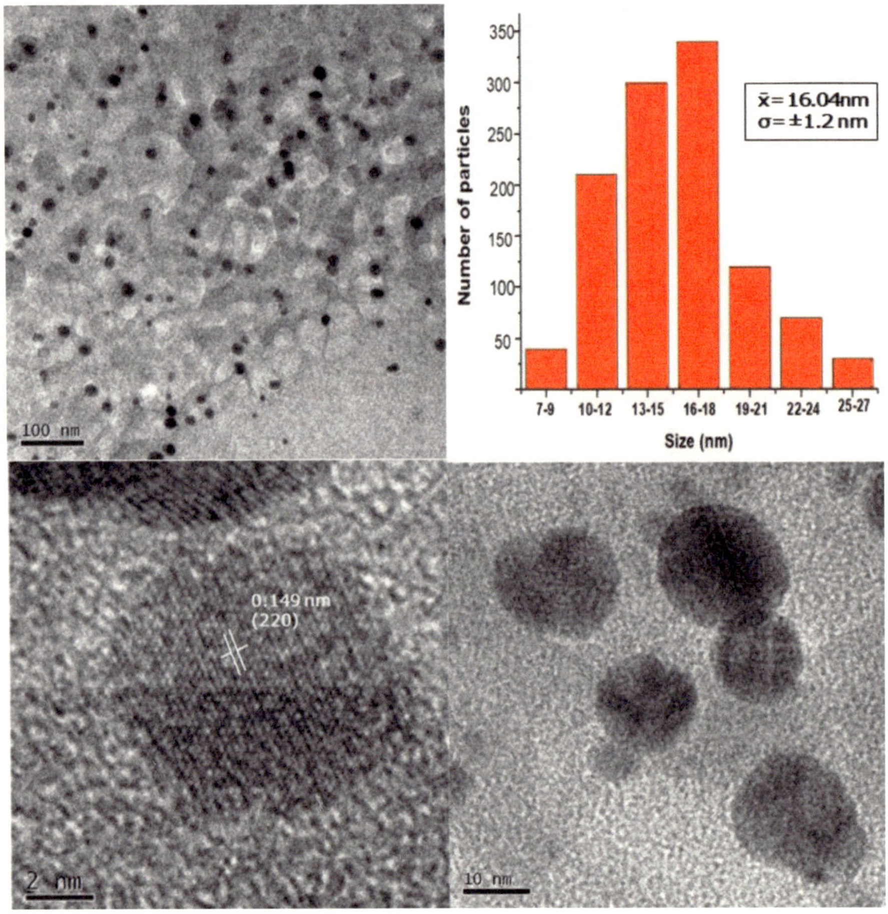

Figure 2.
TEM images show the size distribution and a spherical shape of AgNPs synthesized by the green reducing agent, having a mean diameter of approximately 17 nm.

deviation of 1.2 nm are recognized; in addition, an interplanar distance of 0.149 nm, corresponding to plane (220 nm), was found (**Figure 2**).

Scanning electron microscopy images of catgut embedded with AgNPs (reduced with *Heterotheca inuloides*) at different magnifications are shown in **Figure 3**. Ag nanoparticles of spherical in shape are observed distributed over the fiber surface.

4. Characterization of bioreducing agent of AgNp by infrared spectroscopy

H. inuloides represents a source of chemical compounds with variable structural patterns. Several different types of compounds such as sesquiterpenes, triterpenes, polyphenols, and phytosterols have been isolated from essential oil and organic extracts from various parts, including roots, aerial parts, and flowers. According to the abovementioned, the following characteristic functional groups, 3268 cm^{-1} (–OH), 2942 cm^{-1} (c (sp2)–H), 1584 cm^{-1} (C = O), 1393 cm^{-1} ($-CH_2$), 1258 cm^{-1} ($-CH_3$), 1033 cm^{-1} (CO), and 595 cm^{-1} (CH), were detected (**Figure 4**).

The antimicrobial activity of the infusion using *Heterotheca inuloides* as reducing agent, against *Staphylococcus aureus*, can be seen in **Figure 5**. A well-defined inhibition halo around the disk impregnated with the nanoparticles solution is visible.

DOI: http://dx.doi.org/10.5772/intechopen.89344

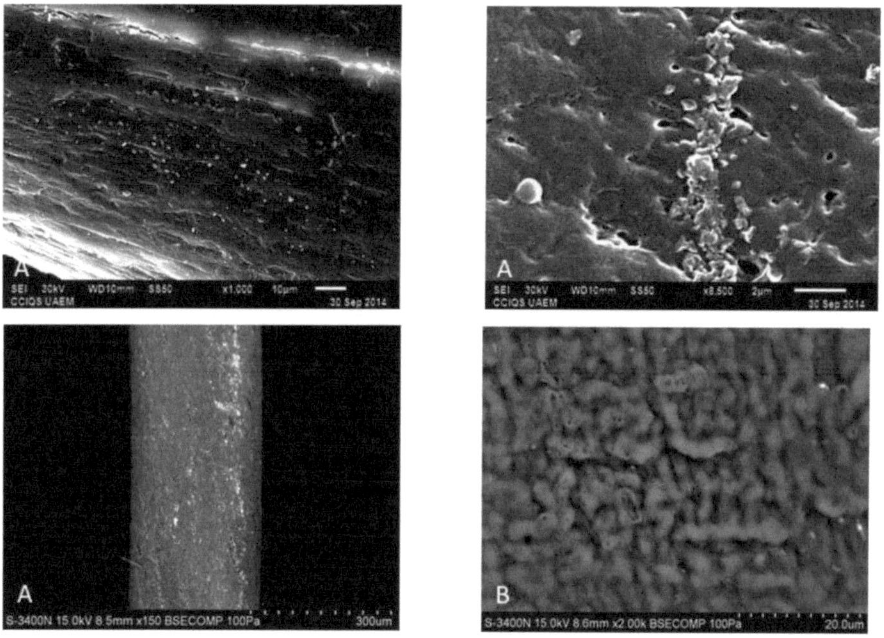

Figure 3.
SEM micrographs showing the catgut suture threads coated with AgNPs synthesized by the green reducing agent. (A) Images revealed that AgNPs were formed on the surface. (B) The micrograph shows a plain catgut suture.

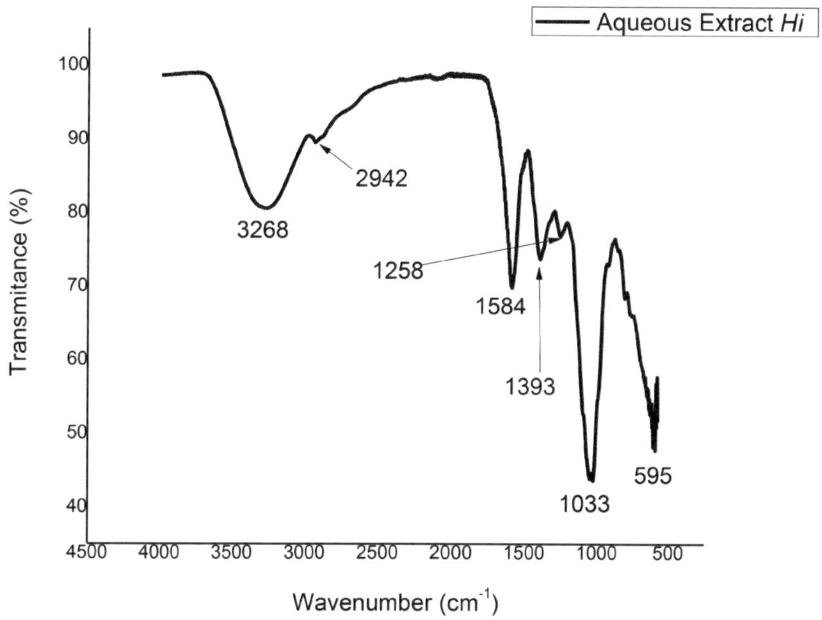

Figure 4.
FTIR spectrum of the Heterotheca inuloides *aqueous extract.*

In **Figure 6**, the antimicrobial activity of catgut against *S. aureus* and *E. coli* is observed. The suture threads were cut into small pieces and put on the Petri dishes. Some suture thread samples were used as blank.

The inhibition zone for both strains is presented in the next tables.

Table 1 shows that the use of *Heterotheca inuloides* to synthesize AgNPs produces an antibacterial effect against both strains, by testing disks. The growth inhibition

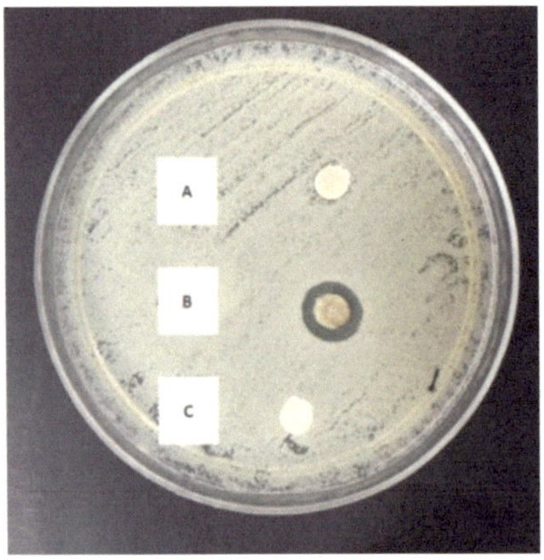

Figure 5.
AgNPs against Staphylococcus aureus: *(A) blank disk, (B) disk containing AgNPs synthesized by* H. inuloides, *and (C) disk with* H. inuloides *infusion as a control.*

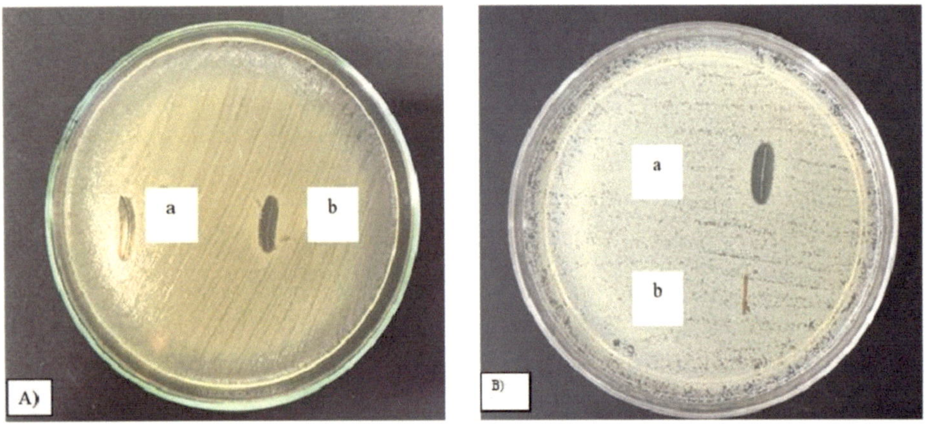

Figure 6.
Antibacterial effect of suture against both strains a and b. (A) Inhibitory halo of catgut suture with AgNPs versus E. coli. (B) Inhibitory halo of catgut suture with AgNPs versus S. aureus. (a) Catgut with AgNPs. (b) Catgut without AgNPs used as a blank.

halo for *S. aureus* was on 3.5 mm average. While for *E. coli*, it was 3.25 mm on average. Without having a statistically significant difference between both strains.

When performing the suture inhibition test for both strains, an average inhibition zone of 3.46 mm for *S. aureus* and an inhibition zone of 2.8 mm for *E. coli* can be seen in **Table 2**, demonstrating a statistically significant difference and a greater zone of inhibition with the use of catgut versus *S. aureus*.

There was no growth inhibition with blank or control disks, neither with catgut blank sutures. All the measurements were replicated three times for each treatment.

5. Discussion

Regarding the UV-Vis results, we can recognize the presence of the characteristic surface plasmonic resonance of silver nanoparticles as other authors have reported to appear from 400 to 500 nm [14].

Other authors have shown that silver nanoparticles with sizes smaller than 50 nm offer high antimicrobial activity [15] that supported on catgut fibers and obtain a

Variable	**Obs**	**Mean**	**Std. Err.**	**Std. Dev.**	**[95% Conf. Interval]**
Sdis~pHi	4	3.5	0.2886751	0.5773503	2.581307–4.418693
Edis~pHi	4	3.25	0.25	0.5	2.454388–4.045612
combined	8	3.375	0.1829813	0.5175492	2.942318–3.807682
diff		0.25	0.3818813		-0.6844299–1.18443
diff = mean(Sdisco_NpHi) – mean(Edisco_NpHi)				t = 0.6547	
Ho: diff = 0				degrees of freedom = 6	
Ha: diff < 0		Ha: diff != 0		Ha: diff > 0	
Pr(T < t) = 0.7315		Pr(|T| > |t|) = 0.5370		Pr(T > t) = 0.2685	

Table 1.
Measures of the zones of inhibition of the disks against S. aureus *and* E. coli.

Two-sample t test with equal variances					
Variable	**Obs**	**Mean**	**Std. Err.**	**Std. Dev.**	**[95% Conf. Interval]**
Scatgu~i	15	3.466667	0.1652319	0.6399405	3.112279–3.821054
Ecatgu~i	15	2.8	0.2225395	0.8618916	2.3227–3.2773
combined	30	3.133333	0.1495844	0.8193072	2.827399–3.439268
diff		0.6666667	0.2771739		0.0989016–1.234432
diff = mean(Scatgut_NpHi) – mean(Ecatgut_NpHi)				t = 2.4052	
Ho: diff = 0				degrees of freedom = 28	
Ha: diff < 0		Ha: diff != 0		Ha: diff > 0	
Pr(T < t) = 0.9885		Pr(|T| > |t|) = 0.0230		Pr(T > t) = 0.0115	

Table 2.
When comparing both strains; there was a mean inhibition zone of 3.4 mm against S. aureus. *While for* E. coli *there was a mean inhibition zone of 2.8 mm representing a statistical significant difference of 0.0230 value.*

double function, in **Figure 3**. The accommodation of the nanoparticles on the surface of the fibers can be observed, and some authors have observed this same behavior [16].

Also, the main functional groups present in the reducing agent are recognized [1], which makes it possible to obtain silver nanoparticles with average sizes of 16.04 nm.

According to the Centers for Disease Control and Prevention, assessment of wound healing after a surgical procedure is important [17]. Infection is the most important and preventable cause of impaired wound healing. Microorganisms can rapidly reach tissues after surgery [18].

A surgical site infection (SSI) surveillance in oral or maxillofacial surgery is necessary because the mouth is widely colonized by highly pathogenic microorganisms. Besides, suture threads are placed in a moist environment [19]. One of the categories to classify the SSI is the complications related to the superficial incision, in which the suture material used may be related [20]. Whether its natural or synthetic origin, the sutures are strange to the body and therefore cause tissue reaction. Any suture may provide an environment conducive to the propagation of infection [21].

In 2002 the Food and Drug Administration (FDA) approved the use of the first suture coated with antimicrobial and triclosan (polyglactin 910-polychlorophenoxyphenol), which has a broad-spectrum activity against Gram-positive and Gram-negative microorganisms [12].

However, the resistance of highly pathogenic microorganisms has been reported as a disadvantage of the use of triclosan [22].

An advanced material for the closure of a wound, with direct drug delivery from the suture to the surgical site, may be the key for the prevention of infections.

Mexico is one of the five megadiverse countries of the world [23]. Pharmaceutical products derived from this plant are widely used, since ancient times, due to the botanical exploration of the Valley of Mexico, one of the most well-known regions from the scientific and botanical aspect [24]. The Ministry of Health has studied its traditional therapeutic use in most regions of the country, using the flower as an infusion and other presentations, [25] searching for the establishment of public policies, and recognizing the importance of epidemiological contributions of popular medicine [26].

Plants have different types of metabolites that can help in the reduction of silver ions. The biological methods using plant leaf extracts have shown great potential for nano-synthesis [27]. Using endemic plants provides many advantages, such as easy accessibility, simplicity, economy, and ecological nature [28–31].

Some of the most important considerations of a green synthetic method are the use of nontoxic chemicals, benign solvents for the environment, and renewable materials [10, 32]. Besides, the process can be carried out at room temperature, and the reaction is completed in a few minutes. Green synthetic methods are simple, environmentally benign, and quite efficient for producing silver nanoparticles [10, 28].

6. Conclusion

We present a totally green approach toward the synthesis and stabilization of metal nanoparticles allowing to obtain an advanced suture that can be effective on oral and maxillofacial surgery, having demonstrated an antibacterial effect versus resistant bacteria. In this study, *Heterotheca inuloides* turns out to be an appropriate reducing agent for coating natural suture threads with AgNPs.

Synthesis of AgNp by eco-friendly reducing agents represents an environmental and economically sustainable process that minimizes the costs and provides the benefits and properties of plants such as *Heterotheca inuloides*.

We believe that this may be an alternative for surgeons, which helps in reducing the indiscriminate use of antibiotic therapy. It also represents an option to use advanced materials that are produced under sustainable conditions, which reduce the impact on the environment.

Conflict of interest

The authors declare that they do not have conflict of interest.

Author details

Saraí C. Guadarrama-Reyes[1], Raúl A. Morales-Luckie[2*], Víctor Sánchez-Mendieta[2], María G. González-Pedroza[2], Edith Lara-Carrillo[3], Ulises Velazquez-Enriquez[3], Victor Toral-Rizo[3] and Rogelio Scougall-Vilchis[3]

1 School of Dentistry, Autonomous University of the State of Mexico, Toluca, State of Mexico, Mexico

2 Joint Center for Research in Sustainable Chemistry UAEM-UNAM (CCIQS), Autonomous University of the State of Mexico, Toluca, State of Mexico, Mexico

3 Center for Research and Advanced Studies in Dentistry, School of Dentistry, Autonomous University of the State of Mexico, Toluca, State of Mexico, Mexico

*Address all correspondence to: rmoralesl@uaemex.mx; ramluckie@gmail.com

References

[1] Rodríguez-Chávez JL et al. Mexican Arnica (*Heterotheca inuloides* Cass. Asteraceae: Astereae): Ethnomedical uses, chemical constituents and biological properties. Journal of Ethnopharmacology. 2017;**195**:39-63

[2] Delgado G et al. Anti-inflammatory constituents from *Heterotheca inuloides*. Journal of Natural Products. 2001;**64**(7):861-864

[3] Coballase-Urrutia E et al. Antioxidant activity of *Heterotheca inuloides* extracts and of some of its metabolites. Toxicology. 2010;**276**(1):41-48

[4] Coballase-Urrutia E et al. Hepatoprotective effect of acetonic and methanolic extracts of *Heterotheca inuloides* against CCl_4-induced toxicity in rats. Experimental and Toxicologic Pathology. 2011;**63**(4):363-370

[5] Rosas-Piñón Y et al. Ethnobotanical survey and antibacterial activity of plants used in the Altiplane region of Mexico for the treatment of oral cavity infections. Journal of Ethnopharmacology. 2012;**141**(3):860-865

[6] World Health Organization. General Guidelines for Methodologies on Research and Evaluation of Traditional Medicine. Geneva: World Health Organization; 2000

[7] Gené RM et al. *Heterotheca inuloides*: Anti-inflammatory and analgesic effect. Journal of Ethnopharmacology. 1998;**60**(2):157-162

[8] Kubo I et al. Antimicrobial agents from *Heterotheca inuloides*. Planta Medica. 1994;**60**(03):218-221

[9] Rodríguez-Chávez JL et al. In vitro activity of 'Mexican Arnica' *Heterotheca inuloides* Cass natural products and some derivatives against *Giardia intestinalis*. Parasitology. 2015;**142**(4):576-584

[10] Roy N et al. Green synthesis of silver nanoparticles: An approach to overcome toxicity. Environmental Toxicology and Pharmacology. 2013;**36**(3):807-812

[11] Jaidev L, Narasimha G. Fungal mediated biosynthesis of silver nanoparticles, characterization and antimicrobial activity. Colloids and Surfaces B: Biointerfaces. 2010;**81**(2):430-433

[12] Granados-Romero JJ et al. Comparación entre sutura recubierta con antibacterial versus cierre tradicional en la incidencia de complicaciones en apendicectomías y colecistectomías laparoscópicas. Revista Mexicana de Cirugía Endoscópica. 2015;**16**(1-4):31-36

[13] Corrêa JM et al. Silver nanoparticles in dental biomaterials. International Journal of Biomaterials. 2015;**2015**:485275

[14] Sharma VK, Yngard RA, Lin Y. Silver nanoparticles: Green synthesis and their antimicrobial activities. Advances in Colloid and Interface Science. 2009;**145**(1-2):83-96

[15] Morales-Luckie RA et al. Synthesis of silver nanoparticles using aqueous extracts of *Heterotheca inuloides* as reducing agent and natural fibers as templates: *Agave lechuguilla* and silk. Materials Science and Engineering: C. 2016;**69**:429-436

[16] Aramwit P et al. Green synthesis of silk sericin-capped silver nanoparticles and their potent anti-bacterial activity. Nanoscale Research Letters. 2014;**9**(1):79

[17] Mangram AJ et al. Guideline for prevention of surgical site infection,

1999. Infection Control and Hospital Epidemiology. 1999;**20**(4):247-280

[18] Bickler SW, Spiegel D. Improving surgical care in low-and middle-income countries: A pivotal role for the World Health Organization. World Journal of Surgery. 2010;**34**(3):386-390

[19] Leknes KN et al. Tissue reactions to sutures in the presence and absence of anti-infective therapy. Journal of Clinical Periodontology. 2005;**32**(2):130-138

[20] Kathju S et al. Chronic surgical site infection due to suture-associated polymicrobial biofilm. Surgical Infections. 2009;**10**(5):457-461

[21] Fantry AJ et al. Deep infections after Syndesmotic fixation with a suture button device. Orthopedics. 2017;**40**(3):e541-e545

[22] Edmiston CE Jr, Daoud FC, Leaper D. Is there an evidence-based argument for embracing an antimicrobial (triclosan)-coated suture technology to reduce the risk for surgical-site infections? A meta-analysis. Surgery. 2013;**154**(1):89-100

[23] Castillo-Juárez I et al. Anti-*Helicobacter pylori* activity of plants used in Mexican traditional medicine for gastrointestinal disorders. Journal of Ethnopharmacology. 2009;**122**(2):402-405

[24] Rzedowski GD, Rzedowski J. Flora Fanerogámica del Valle de México. 1a reimp ed. Pátzcuaro, Michoacán: Instituto de Ecología, AC y Comisión Nacional para el Conocimiento y Usos de la Biodiversidad; 2005

[25] Hernández-Cruz AS. Manual para el manejo sustentable de plantas medicinales y elaboración de productos derivados. Instituto Nacional de Desarrollo Social, Indesol. México. 2014:63. Folio CS-09-F-DI-308-14

[26] Almaguer G. Programa de Trabajo 2001-2006 de la Dirección de Medicina Tradicional. Coordinación de Salud para los Pueblos Indígenas. México, DF: Secretaría de Salud; 2001. Manuscrito

[27] Shinde N, Lokhande A, Lokhande C. A green synthesis method for large area silver thin film containing nanoparticles. Journal of Photochemistry and Photobiology B: Biology. 2014;**136**:19-25

[28] Behravan M et al. Facile green synthesis of silver nanoparticles using *Berberis vulgaris* leaf and root aqueous extract and its antibacterial activity. International Journal of Biological Macromolecules. 2019;**124**:148-154

[29] Anand KKH, Mandal BK. Activity study of biogenic spherical silver nanoparticles towards microbes and oxidants. Spectrochimica Acta Part A: Molecular and Biomolecular Spectroscopy. 2015;**135**:639-645

[30] Bindhu M, Umadevi M. Antibacterial and catalytic activities of green synthesized silver nanoparticles. Spectrochimica Acta Part A: Molecular and Biomolecular Spectroscopy. 2015;**135**:373-378

[31] Ulug B et al. Role of irradiation in the green synthesis of silver nanoparticles mediated by fig (*Ficus carica*) leaf extract. Spectrochimica Acta Part A: Molecular and Biomolecular Spectroscopy. 2015;**135**:153-161

[32] Raveendran P, Fu J, Wallen SL. Completely "green" synthesis and stabilization of metal nanoparticles. Journal of the American Chemical Society. 2003;**125**(46):13940-13941

Chapter 2

Preparation of Nanoparticles

Takalani Cele

Abstract

Innovative developments of science and engineering have progressed very fast toward the synthesis of nanomaterials to achieve unique properties that are not the same as the properties of the bulk materials. The particle reveals interesting properties at the dimension below 100 nm, mostly from two physical effects. The two physical effects are the quantization of electronic states apparent leading to very sensitive size-dependent effects such as optical and magnetic properties and the high surface-to-volume ratio modifies the thermal, mechanical, and chemical properties of materials. The nanoparticles' unique physical and chemical properties render them most appropriate for a number of specialist applications.

Keywords: nanoparticles, physical methods, chemical methods, synthesis, metal nanoparticles

1. Introduction

Several methods have been developed to produce metal nanoparticles. Two synthesis approaches have been identified that is top-down and bottom-up approach. Top-down methods comprise of milling, lithography, and repeated quenching. This approach does not have good control of the particle size and structure. Bottom-up method is the approach that is mostly used by scientists in the synthesis of nanoparticles as it involves building up a material from bottom: atom-by-atom, molecule-by-molecule, and cluster-by-cluster [1, 2]. Several chemical routes have been identified to synthesize the colloidal metal nanoparticles from different precursors using chemical reductants in solvents (aqueous and nonaqueous). The chemical routes that have been studied for various applications include electrochemical method [3], sonochemical method [4], radiolytic [5] and photochemical [6] method.

2. Methods of synthesis of metal nanoparticles

2.1 Chemical methods

2.1.1 The polyol method

The Polyol method is a chemical method for the synthesis of nanoparticles.

This method uses nonaqueous liquid (polyol) as a solvent and reducing agent. The nonaqueous solvents that are used in this method have an advantage of minimizing surface oxidation and agglomeration. This method allows flexibility on controlling of size, texture, and shape of nanoparticles. Polyol method can also be used in producing nanoparticles in large scale [7].

The polyol process can be taken as a sol-gel method in the synthesis of oxide, if the synthesis is conducted at moderately increased temperature with accurate particle growth control [8]. There are several reports that have studied the synthesis of oxide sub-micrometer particles and these include Y_2O_3, V_xO_y,Mn_3O_4, ZnO, $CoTiO_3$, SnO_2, PbO, and TiO_2 [9–16].

The solvent that is mostly used in polyol method in metal oxide nanoparticles synthesis is ethylene glycol because of its strong reducing capability, high dielectric constant, and high boiling point. Ethylene glycol is also used as a crosslinking reagent to link with metal ion to form metal glycolate leading to oligomerization [17]. It has been reported that as-synthesized glycolate precursors can be converted to their more common metal oxide derivatives when calcined in air, while maintaining the original precursor morphology [8].

The polyol synthesis process has also been used for the synthesis of bimetallic alloys and core-shell nanoparticles [18–20]. Yang and co-workers used polyol method to produce icosahedral and cubic gold particles on the order of 100–300 nm by careful regulation of the growth rate for each crystallographic direction [21]. Xia and co-workers reported the production of controlled morphologies such as nanocubes and nanowires by controlling the molar ratio between silver nitrate and *PVP* [22].

2.1.2 Microemulsions

An emulsion is a liquid in liquid dispersion. A solution of polymers can produce emulsions as it is liquid. Emulsions are divided according to the size of droplet, i.e., macro-emulsions, mini-emulsions, and micro-emulsions [23].

Micro-emulsion synthesis method is widely used for the production of inorganic nanoparticles [24]. When oil and water are mixed, they separate into two phases as they are immiscible [25]. The energy input is required to mix the two phases to create water-oil.

An attempt to combine the two phases requires energy input that would establish water-oil connection replacing the water-water/oil-oil contacts. The interfacial tension between bulk oil and water can be as high as 30–50 dynes/cm and this can be avoided by using surfactants (surface-active molecules). Surfactants contain hydrophilic (water-loving) and lipophilic (oil-loving) groups [26]. The interface can be aligned and established between oil and water by reducing the interfacial tension if there are enough surfactant molecules.

The preparation procedure of metallic nanoparticles in water in oil microemulsion commonly consists of mixing of two microemulsions containing metal salt and a reducing agent, respectively as shown in **Figure 1**.

Brownian motion is formed after the exchange of reactants (collision) between micelles that happens after mixing two microemulsions. Good collisions result into coalescence, fusion, and mixing well of the reactants. Metal nuclei are formed from the reaction between solubilizates. Bönnemann et al. reported the formation of zerovalent metal atoms at nucleation stage from reducing a metal salt, which can collide with additional metal ions, metal atoms, or clusters to form an irreversible seed of stable metal nuclei [28].

The growth stage happens around the nucleation point, where successful collision occurs between a reverse micelle moving a nucleus and another one moving the product monomers with the arrival of more reactants due to intermicellar exchange. The morphology and size of nanoparticles are based on the size and shape of the nanodroplets and the type of the surfactant. The surfactant is usually used to stabilize the particle and protect them from proceeding to grow [28].

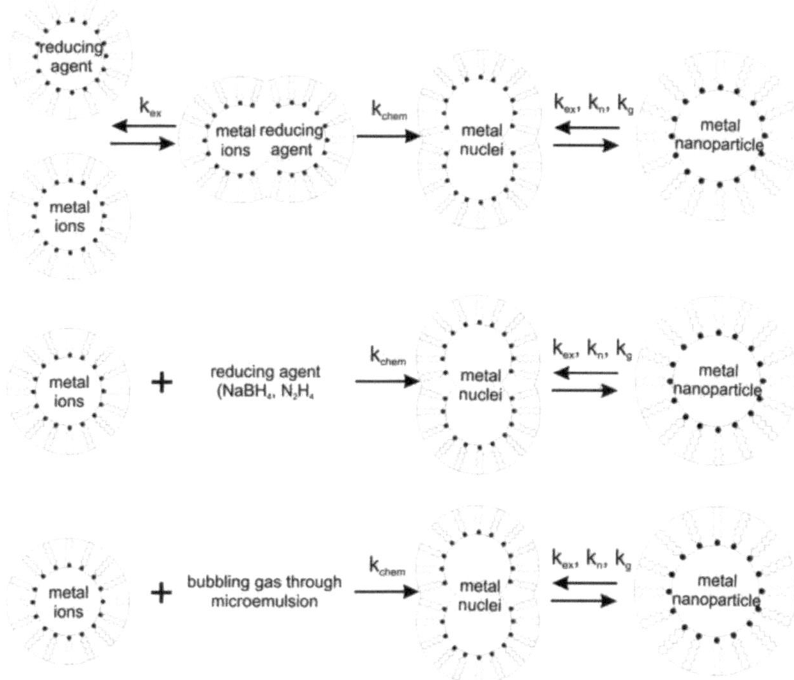

Figure 1.
Schematic illustration of nanoparticles preparation using microemulsion techniques: Particle formation steps. K_{chem} is the rate constant for chemical reaction, k_{ex} is the rate constant for intermicellar exchange dynamics, k_n is the rate constant for nucleation, and k_g is the rate constant for particle growth [27].

Wongwailikhit et al. reported the formation of iron (III) oxide, Fe_2O_3 using water in oil microemulsion by combining the required amount of H_2O in a stock solution of Sodium Bis (2-Ethylhexyl) Sulfosuccinate (AOT) in *n*-heptane. The solution was left overnight, then the concentrated Hydroxylamine (NH_2OH) and $FeCl_3$ were mixed into the water in oil microemulsion. Suspension of Fe_2O_3 was filtered and washed with 95% ethanol and dried at 300°C for 3 h. The product was spherical, monodisperse nanoparticles with diameter of about 50 nm. The size of particles depended on the water content in microemulsion system. The increase of particles size was achieved with increasing the water fraction in water in oil microemulsion [29].

Sarkar et al. reported the formation of pure monodispersed zinc oxide nanoparticles of different shapes. Microemulsion was composed of cyclohexane, Triton X-100 as surfactant, hexanol as cosurfactant and aqueous solution of zinc nitrate or ammonium hydroxide/sodium hydroxide complex. The molar ratio of TX-100 to hexanol was maintained at 1:4. The microemulsion containing ammonium hydroxide/sodium hydroxide was added to microemulsion containing zinc nitrate and stirred. The nanoparticles were then separated by centrifuging at 15,000 rpm for 1 h. The particles were washed with distilled water and alcohol and dried at 50°C for 12 h [30].

Maitra was the first to establish Chitosan nanoparticles by microemulsion technique. Chitosan nanoparticles were prepared in the aqueous core of reverse micellar droplets and crosslinked through glutaraldehyde. Surfactant dissolved in N-hexane was also used with chitosan in acetic acid and glutaraldehyde was added in the surfactant at room temperature. The mixture was stirred continuously and nanoparticles were produced [31].

2.1.3 Thermal decomposition

Thermal decomposition also known as thermolysis is a chemical decomposition that is caused by heat. In this method, the heat is required to break chemical bonds in the compound undergoing decomposition and the reaction is endothermic. If decomposition is sufficiently exothermic, a positive feedback loop is created producing thermal runaway [32].

Arshad et al. reported on thermal decomposition of metal complexes of type MLX_2 [M = Co (II), Cu (II), Zn (II), and Cd (II);L = DIE; X = NO_3^{1-}] by TG-DTA-DTG techniques in air atmosphere. They synthesized nitrate complexes of transition metals with 1,2-diimidazoloethane (DIE) of the general formula M(DIE)$(NO_3)_2$. The study was conducted by thermoanalytical techniques in static air atmosphere to study the thermal behavior of these complexes and to determine their mode of decomposition. The complexes and ligands decomposed in a two-step process when heated to 740°. Above 740°, the residue was found to correspond with metal oxide. The thermal stability of the complexes increases in the following series: Co(II) < Cu(II) < Zn(II) < Cd(II) [33].

Patil et al. studied infrared spectra and thermal decompositions of metal acetates and dicarboxylates. The study was done to determine the metal-acetate bonding and the thermal decomposition of lead, copper, and rare earth acetates was studied by means of thermogravimetric analysis and differential thermal analysis. The investigations on decomposition products yielded good results [34].

George et al. reported on the mechanism of thermal decomposition of *n*-Buty l (tri-*n*-butylphosphine) copper (I). This study provided the first easily interpretable example in which succeeding reaction of a metal hydride and its parent metal alkyl was found to be vital in determining the products of a thermal decomposition [35].

Thermal decomposition of bismuth and silver carboxylates was investigated by means of TG, DSC, mass spectrometry, X-ray analysis, and electron microscopy Logvinenko et al. [36]. Non-isothermal thermogravimetric data were used for kinetic studies. All decomposition processes had multi-step character [36].

Ewell et al. investigated nearly pure talc both unheated and after heating at various temperatures ranging up to 1,435°C. The research included the measurement of heat effects, weight losses, and changes in true specific gravity occurring on heating talc. There was no change in the crystal structure of the talc heated up to 800°C. At the temperature between 800 and 8400°C, the talc decomposed to enstatite, amorphous silica, and water vapor. At the temperature approximately 1,200°C, the enstatite steadily changed to clinoenstatite and the amorphous silica changed to cristobalite approximately 1,300°C, giving clinoenstatite and cristobalite as end products [37].

2.1.4 Electrochemical synthesis

Electrochemical synthesis is the synthesis of chemical compounds in an electrochemical cell. The main advantage of electrochemical synthesis over an ordinary chemical reaction is rejection of the potential wasteful alternative half-reaction and the ability to accurately tune the preferred potential [38].

Electrochemical synthesis of silver nanoparticles has been extensively studied in the previous years. The method of electrochemical that was used was based on the dissolution of a metallic anode in an aprotic solvent. The silver nanoparticles that were produced by electroreduction of anodically solved silver ions in acetonitrile containing tetrabutylammonium ranged from 2 to 7 nm. The particle size was obtained by varying the current density. Different types of counter electrodes were used to study the effect of the different electrochemical parameters on the end particle size. The UV-Vis spectra showed the presence of two different silver clusters [39].

Dobre et al. also reported on the electrochemical synthesis of colloidal silver solutions using "sacrificial anode" technique conducted with a home-built current pulse generator with alternating polarity and a stirrer. Poly (N-vinyl-2-pyrrolidone) (PVP) and sodium lauryl sulfate (Na-LS) were used as a stabilizer and co-stabilizer, correspondingly. Spherical Ag particles with the size approximately 10–55 nm were synthesized. The UV/Vis spectra showed the absorption band at 420 nm, which is the evidence of the presence of Ag nanoparticles. The zeta potential values between −17 and −35 mV suggested a presence of particles covered by stabilizer with a slight agglomeration [40].

More research was done on the electrochemical synthesis of silver nanoparticles in aqueous poly (vinyl alcohol) solution (PVA). PVA is a low price widely used synthetic polymer with properties such as nontoxicity, water solubility, biocompatibility, biodegradability, and excellent mechanical properties. The experiment was conducted at a constant current density of 25 mA cm^{-2} for a synthesis time of 10 min. Silver nanoparticles with an average diameter of 15 ± 9 nm were obtained [41].

The electrochemical synthesis of red fluorescent Silicon (Si) nanoparticles stabilized with styrene. Si nanoparticles emit fluorescence under UV excitation, which is great for optics applications, etc. It was found that the liberated silicon particles in ethanol solution interact with styrene, which resulted in the substitution of Si-H bonds with those of Si-C. The developed styrene-coated Si nanoparticles exhibited a stable, bright, red fluorescence under excitation with a 365 nm UV light, and resulted into approximately 100 mg per Si wafer with a synthesis time of 2 h [42].

More investigations were done on the preparation of long-lived silver nanoparticles in aqueous solutions and silver powders using electrochemical method. The produced silver nanoparticles had a size distribution ranging from 2 to 20 nm and the nanoparticles remained stable for more than 7 years. Silver crystals containing agglomerated silver nanoparticles with sizes below 40 nm was found growing on the surface of the cathode [43].

The research was conducted on using electrochemical method to synthesize highly pure silver nanoparticles. This method was used as it is one-step less expensive procedure and easy to control at room temperature and it does not use dangerous chemicals. The experimental setup brought up the oxidation of the anode and reduction of the cathode. The silver nanoparticles synthesized were spherical and had a particle size below 50 nm [44].

Islam et al. explored on the synthesis of platinum nanoparticles by electrochemical deposition method. The particle size was controlled by varying electrolysis parameters and homogeneity of platinum particles was improved by varying the composition of electrolytic solutions. Platinum nanoparticles were deposited on electrode surfaces and the particle sizes were found to be larger than 10 nm and had wide particle size distribution [45].

2.2 Physical methods

2.2.1 Plasma

Plasma method is another method that is used to produce nanoparticles. The plasma is generated by radio frequency (RF) heating coils. The initial metal is enclosed in a pestle and the pestle is enclosed in an evacuated chamber. The metal is then heated above its evaporation point by high voltage RF coils wrapped around the evacuated chamber. The gas that is used in the procedure is Helium (He), which forms a high-temperature plasma in the region of the coils after flowing into the system. The metal vapor nucleates on the helium gas atoms and diffuses up to a cold collector rod, this is where nanoparticles are collected and they are passivated by oxygen gas (**Figures 2** and **3**) [46].

Classification of plasma methods based on the feeding materials to reactor and also the heating source (electrodeless/ electrode containing), see (**Figures 2** and **3**).

2.2.2 Chemical vapor deposition

The chemical vapor deposition method (CVD) involves a chemical reaction. CVD procedure is mostly used in semiconductor manufacturing for depositing thin films of different materials. The method involves one or more volatile precursors, the substrate is exposed to those precursors that decompose on it and form the desired deposit. The vaporized precursors are inserted into a CVD reactor and adsorb onto a substance being placed at high temperature. The molecules that get adsorbed react with other molecules or decompose to form crystals. The three steps in CVD method are:

1. Reactants are transported on the growth surface by a boundary layer.

2. Chemical reactions occur on the growth surface.

3. By products produced by the gas-phase reaction has to be removed from the surface. Homogeneous nucleation occurs in gas phase and heterogeneous nucleation happens in a substrate.

The CVD method can synthesize ultrafine particles of less than 1 μm by the chemical reaction taking place in the gaseous phase. The reaction can be controlled to produce nanoparticles of size ranging from 10 to 100 nm [46, 47].

2.2.3 Microwave irradiation

Microwave irradiation is a synthesis method that has been widely used in the synthesis of organic, inorganic, and inorganic–organic hybrid materials because of its well-known advantages over conventional synthetic routes [48].

A research was conducted on a rapid and efficient oxidation of organic compounds in microwave condition with new phase transfer oxidative agent: CTAMABC. CTMABC (1 mmole) was suspended in acetonitrile (2 ml) and an alcohol (l mmole in 0.5–1.5 ml of acetonitrile) was quickly added at room temperature and the resulting mixture was stirred vigorously. The mixture was then irradiated by microwave radiation (3.67 GHz, 300 W). The solution became homogeneous for a short time before the black-brown reduced reagent precipitated. Thin layer chromatography (TLC) and UV/VIS spectrophotometer (at 352 nm) were used to monitor the progress of reactions [49].

In another experiment conducted by Sahoo Biswa Mohan et al., o-Phenylenediamine (1.08 g, 0.01 mole) and anthranillic acid (1.37 g, 0.01mole) were dissolved in ethanol (15 ml). And K_2CO_3 was added to a mixture and the reaction mixture was put in microwave oven and refluxed at power (140 Watt) for 10 min. TLC was used to monitor the reaction. After the reaction was complete, ethanol was removed by distillation process and the residue was poured into crushed ice. Then the reaction was made alkaline by using 10% NaOH to get the solid product. The product was filtered, dried, and recrystallized from ethanol [50].

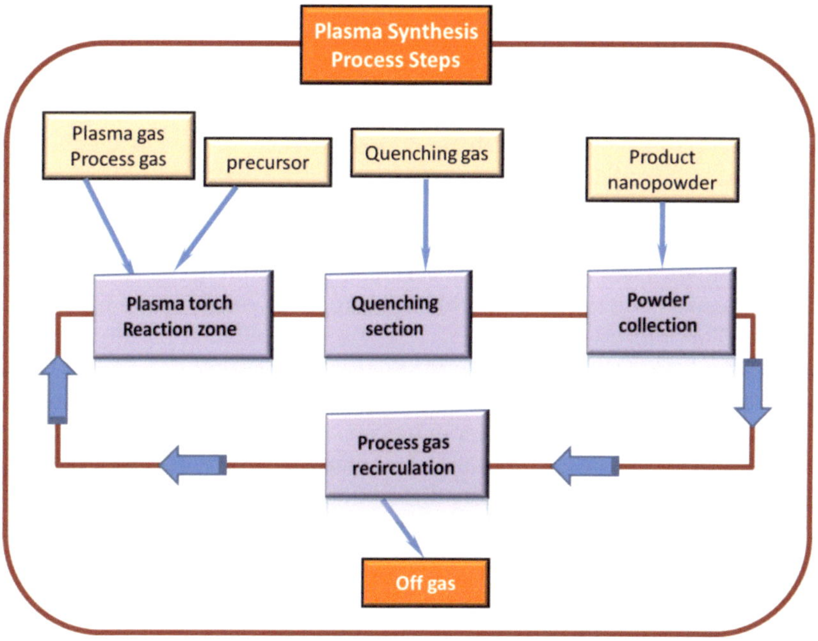

Figure 2.
Flow diagram for production plant based on plasma burners. The recirculation system is of special importance in the case of expensive reaction or carrier gases.

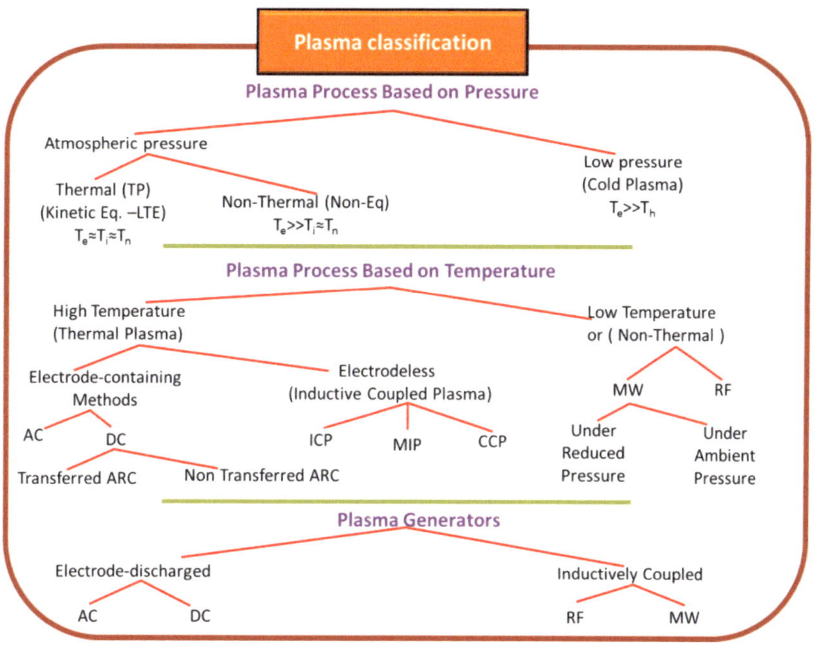

Figure 3.
Different plasma classification.

Sahoo Biswa Mohan et al. conducted another experiment where N-(2-(1H-benzo[d]imidazol-2-yl) phenyl) acetamide (2.51 g, 0.01 mole) was dissolved in ethanol (30 ml) and various aromatic aldehydes (0.01 mole) were taken and then an aqueous solution of KOH (2%, 5 ml) added to it. The reaction was then put in a microwave oven and refluxed at power (210 Watt) for 10–20 min. The excess solvent was removed by vacuum distillation and then poured into crushed ice and acidified with dilute HCl. The product was filtered, dried, and recrystallized from ethanol [51].

Microwave-assisted organic synthesis has been widely used due to enhanced reaction rates, higher yields, improved purity, ease of work up after the reaction and eco-friendly reaction conditions compared to the conventional methods. In above experiments, microwave irradiated synthesis of chalcone was carried out to get higher yield with less reaction time period as compared to conventional method.

The synthesized benzimidazolyl chalcone produces yield around 60% (conventional) and 80% (microwave) [52].

Another study was conducted to synthesize silver nanoparticles (AgNPs) in aqueous medium by a simple, efficient, and economic microwave-assisted synthetic route using hexamine as the reducing agent and the biopolymer pectin as stabilizer. The synthesized AgNPs were characterized by UV-VIS, Spectroscopy, Energy dispersive X-ray (EDX), X-ray diffraction (XRD), and Transmission electron microscopy (TEM) techniques. The nanoparticles were found to be spherical shape with an average diameter of 18.84 nm. The rate of reaction was found to increase with increasing temperature and the activation energy was found to be 47.3 kJ mol^{-1} [53].

ZnS nanoparticles were synthesized by microwave-assisted irradiation method. The produced ZnS nanoparticles were characterized by XRD, SEM, and UV-Vis spectroscopy. The average size of the nanocrystallites was measured by Debye-Scherrer formula as per the XRD spectrum, and there were found to be approximately 6 nm [54].

2.2.4 Pulsed laser method

Pulsed laser method is a method that is mostly used in the synthesis of silver nanoparticles, at a high rate of production of 3 gm/min. Silver nitrate solution and a reducing agent are poured into a blender-like device. The device is composed of a solid disc that rotates with the solution. The disc is exposed to pulses from a laser beam to create hot spots on the surface of the disc. Hot spots are where the silver nitrate reacts with reducing agent to produce silver particles that can be separated by centrifuge. The particle size is controlled by the energy of the laser and angular velocity of the disc [46] (**Figures 4** and **5**).

2.2.5 Sonochemical reduction

Sonochemical method has been studied in the synthesis of metal nanoparticles. The synthesis of different types of metal nanoparticles has been studied by use of the sonochemical reduction of the corresponding metal ions. The sonochemical reduction of MnO^{4-}, Au^{3+}, Au^{+}, and Pd^{2+} in the absence and presence of organic additives were investigated in relation to the synthesis of size and shape controlled metal nanoparticles. The rates of reduction were controlled to control the size and shape of metal nanoparticles. The size of the Au nanoparticles formed from the sonochemical reduction of Au^{3+} was controlled in the presence of an organic stabilizer citric acid [55].

Obreja et al. conducted a study on alcoholic reduction platinum nanoparticles synthesis by sonochemical reduction. H_2PtCl_6 was reduced with methanol, ethanol, and propanol working as solvents and reducing agents, in the presence of capping polymers such as chitosan, polyethylene glycol, and poly (amidehydroxyurethane). The produced nanoparticles size was found to be approximately 3 nm [56].

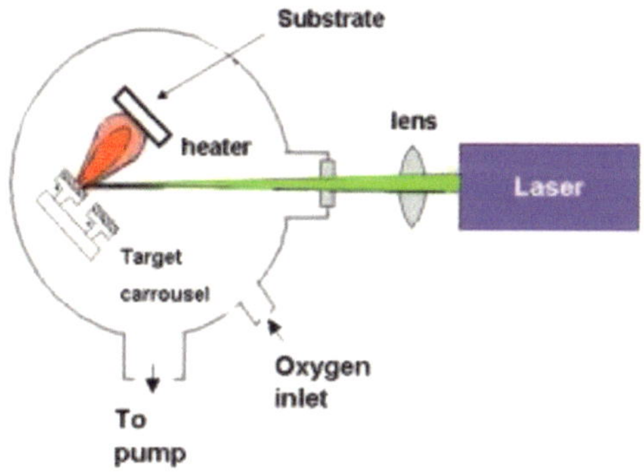

Figure 4.
Synthesis of nanoparticles using a pulsed laser method [46].

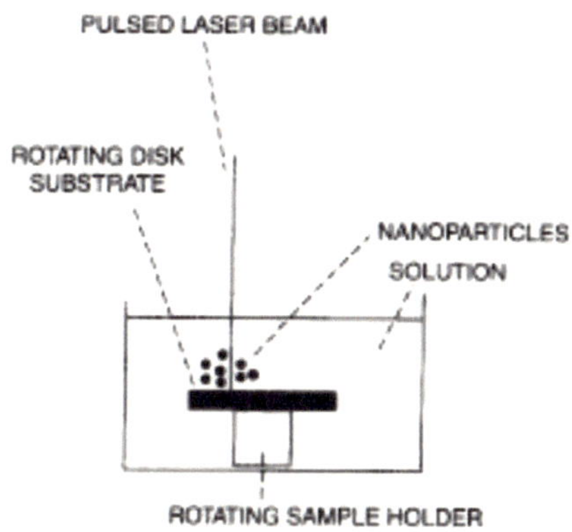

Figure 5.
Apparatus to produce silver nanoparticles using a pulsed laser beam that makes hot spots on the surface of a rotating disk [54].

2.2.6 Gamma radiation

Gamma radiation is the preferred method for metallic nanoparticles synthesis because it is reproducible, may control the shape of the particles yields monodisperse metallic nanoparticles, is easy, cheap, and use less toxins precursors: in water or solvents such as ethanol, it uses the least number of reagents, it uses a reaction temperature close to room temperature with as few synthetic steps as possible (one-pot reaction) and minimizing the quantities of generated by-products and waste [57].

The radiolytic reduction has been proven to be a powerful tool to fabricate monosized and highly dispersed metallic clusters [58]. The primary effects of the interaction of high-energy gamma photons with a solution of metal ions are the excitation and the ionization of the solvent [59]. The different reactions that are observed are well explained in the paper by Abidi and Remita. In particular, water can be produce upon irradiation of a series of reducing and oxidizing agents as shown in the following equation.

$$H_2O \Rightarrow e^-_{aq}, H_3O+, H\bullet, OH\bullet, H_2, H_2O_2 \quad (1)$$

For the production of metallic nanoparticles from metallic salt solutions, the reducing agents e^-_{aq} and H• are the cornerstones of the process. Unfortunately, the production of hydroxyl radicals OH• hampers the efficiency unless some specific hydroxyl scavengers are used. Among them, isopropanol is frequently used [60].

This technique has been widely used so far to produce solutions of MNP primarily gold and silver that were further investigated by UV-Visible spectroscopy with the aim to analyze their plasmonic absorption band. A wealth of literature can be found on this topic [61, 62]. Additionally, γ rays irradiation was also used to trap MNP inside polymers or inside porous frameworks like mesoporous silica for instance [63–65].

3. Conclusions

Nanoparticles have gained significant interest due to their unique chemical and physical properties and are applicable to diverse areas. Various methods of preparation of nanoparticles have been developed and they are suitable for synthesis of nanoparticles in different sizes and shapes. The methods that were discussed include gamma irradiation, chemical reduction photochemical method, thermal decomposition, and microwave irradiation among others.

Acknowledgements

The author is particularly grateful to National Research Foundation, University of South Africa, iThemba LABS and L'Oreal For Women in Science for their support and funding.

Author details

Takalani Cele
University of South Africa, Pretoria, South Africa

*Address all correspondence to: tmadima@yahoo.com

References

[1] Adlim A. Review: Preparations and application of metal nanoparticles. Indonesian Journal of Chemistry. 2006;**6**(1):1-10

[2] Biswal J. A Study on Synthesis of Silver and Gold Nanoparticles by Employing Gamma Radiation, Their Characterization and Applications. Homi Bhabha National Institute; Department of Chemical Sciences; 2012. Available from: http://hdl.handle.net/10603/11458

[3] Hirsch T, Zharnikov M, Shaporenko A, Stahl J, Weiss D, Wolfbeis OS, et al. Size-controlled electrochemical synthesis of metal nano particles on monomolecular templates. Angewandte Chemie International Edition. 2005;**44**:6775-6778

[4] Okitsu K, Mizukoshi Y, Yamamoto TA, Maeda Y, Nagata Y. Sonochemical synthesis of gold nanoparticles on chitosan. Materials Letters. 2007;**61**:3429

[5] Gutierrez M, Henglein A. Formation of colloidal silver by "push-pull" reduction of silver(1+). The Journal of Physical Chemistry. 1993;**97**(44):11368

[6] Sato T, Onaka H, Yonezawa Y. Sensitized photo reduction of silver ions in the presence of acetophenone. Journal of Photochemistry and Photobiology A: Chemistry. 1999;**127**:83-87

[7] Meshesha BT, Barrabés N, Medina F, Sueiras JE. Polyol mediated synthesis and characterization of Cu nanoparticles: Effect of 1-hexadecylamine as stabilizing agent. In: Proceedings of the 1st WSEAS International Conference on Nanotechnology (NANOTECHNOLOGY'09); 2009

[8] Caroline Q, Bernard S, Miele P. Synthesis of praseodymium oxide nanoparticles. In: Nanomaterials and Nanotechnology. Vol. 4. No. Godište 2014; 2014. p. 1+

[9] Feldmann C, Merikhi J. Synthesis and characterization of rod-like Y_2O_3 and Y_2O_3:Eu^{3+}. Journal of Materials Science. 2003;**38**:1731-1735

[10] Ragupathy P, Shivakumara S, Vasan HN, Munichandraiah N. Preparation of nanostrip V_2O_5 by the polyol method and its electrochemical characterization as cathode material for rechargeable lithium batteries. Journal of Physical Chemistry C. 2008;**112**:16700-16707

[11] Sicard L, Le Meins JM, Méthivier C, Herbst F, Ammar S. Polyol synthesis and magnetic study of Mn_3O_4 nanocrystals of tunable size. Journal of Magnetism and Magnetic Materials. 2010;**322**:2634-2640

[12] Lee S, Jeong S, Kim D, Hwang S, Jeon M, Moon J. ZnO nanoparticles with controlled shapes and sizes prepared using a simple polyol synthesis. Superlattices and Microstructures. 2008;**43**:330-339

[13] Siemons M, Simon U. Gas sensing properties of volume-doped $CoTiO_3$ synthesized via polyol method. Sensors and Actuators B: Chemical. 2007;**126**:595-603

[14] Jiang XC, Wang YL, Herricks T, Xia YN. Ethylene glycol-mediated synthesis of metal oxide nanowires. Journal of Materials Chemistry. 2004;**14**:695-703

[15] Flores-Gonzalez MA, Ledoux G, Roux S, Lebboua K, Perriat P, Tillement O. Preparing nanometer scaled Tb-doped Y_2O_3 luminescent powders by the polyol method. Journal of Solid State Chemistry. 2005;**178**:989-997

[16] Jiang XC, Herricks T, Xia YN. Monodispersed spherical colloids of titania: Synthesis, characterization, and crystallization. Advanced Materials. 2003;**15**:1205-1209

[17] Sun YG, Yin YD, Mayers BT, Herricks T, Xia YN. Uniform silver nanowires synthesis by reducing $AgNO_3$ with ethylene glycol in the presence of seeds and poly(vinyl pyrrolidone). Chemistry of Materials. 2002;**14**:4736-4745

[18] Bonet F, Grugeon S, Dupont L, Herrera Urbina R, Guery C, Tarascon J. Synthesis and characterization of bimetallic Ni–Cu particles. Journal of Solid State Chemistry. 2003;**172**:111

[19] Tekaia-Elhsissen K, Bonet F, Silvert P-Y, Herrera-Urbina R. Finely divided platinum–gold alloy powders prepared in ethylene glycol. Journal of Alloys and Compounds. 1999;**292**(1-2):96-99

[20] Garcia-Gutierrez D, Gutierrez-Wing CE, Giovanetti L, Ramallo-Lopez JM, Requejo FG, Jose-Yacaman M. Temperature effect on the synthesis of Au-Pt bimetallic nanoparticles. The Journal of Physical Chemistry B. 2005;**109**:3813-3821

[21] Kim F, Connor S, Song H, Kuykendall T, Yang P. Platonic gold nanocrystals. Angewandte Chemie, International Edition. 2004;**43**(28):3673-3677

[22] Wiley B, Sun Y, Mayers B, Xia Y. Shaped-controlled synthesis of metal nanostructure: The case of silver. Chemistry—A European Journal. 2005;**11**:454-463

[23] Tauer K. MPI Colloids and Interfaces, Emulsions Part 1, Am Mühlenberg, D-14476 Golm, Germany

[24] Yu D, Chu Y, Dong LH, Zhuo YJ. Controllable synthesis of $CaCO_3$ micro/nanocrystals with different morphologies in microemulsion. Chemical Research in Chinese Universities. 2010;**26**:678

[25] Capek I. Radical polymerization of polar unsaturated monomers in direct microemulsion systems. Advances in Colloid and Interface Science. 1999;**80**(2):85-149

[26] Holmberg K. Handbook of Applied Surface and Colloid Chemistry. Chichester, New York: Wiley; 2002

[27] Zielińska-Jurek A, Reszczyńska J, Grabowska E, Zaleska A. Nanoparticles Preparation Using Microemulsion Systems. Poland: Gdansk University of Technology; 2011

[28] Bönnemann H, Richards RM. Nanoscopic metal particles - synthetic methods and potential applications. Carboxylates, physical basis of radiation-related technologies. European Journal of Inorganic Chemistry. 2001;**2001**(10):2455-2480

[29] Wongwailikhit K, Horwongsakul S. The preparation of iron (III) oxide nanoparticles using W/O microemulsion. Materials Letters. 2011;**65**:2820-2822

[30] Sarkar D, Tikku S, Thapar V, Srinivasa RS, Khilar KC. Formation of zinc oxide nanoparticles of different shapes in water-in-oil microemulsion. Colloids and Surfaces A: Physicochemical and Engineering Aspects. 2011;**381**:123-129

[31] Maitra AN, et al. Process for the preparation of highly monodispersed hydrophilic polymeric nanoparticles of size less than 100 nm. US Patent 5874111; 1999

[32] https://en.wikipedia.org/wiki/Thermal_decomposition [Accessed: April 2016]

[33] Arshad M, Rehman S, Quresh AH, Masud K, Arif M, Saeed A, et al.

Thermal decomposition of metal complexes of type MLX_2 (M = Co(II), Cu(II), Zn(II), and Cd(II); L = DIE; X = NO_3^{1-}) by TG-DTA-DTG techniques in air atmosphere. Turkish Journal of Chemistry. 2008;**32**:593-604

[34] Patil KC, Chandrashekhmar GV, George MV, Rao CNR. Infrared spectra and thermal decompositions of metal acetates and dicarboxylates. Canadian Journal of Chemistry. 1968;**46**:257

[35] Whitesides GM, Stedronsky ER, Casey CP, Filippo J. Mechanism of thermal decomposition of *n*-buty l (tri-*n*-buty lphosphine) coppe r(I). Journal of the American Chemical Society. 1970;**92**:1426

[36] Logvinenko V, Mikhailov YV, Polunina OS, Mikhailov K, Minina AV, Yukhin YM. Pecularities of the thermal decomposition of bismuth and silver carboxylates. In: Proceedings, Logvinenko2006PecularitiesOT; 2006

[37] Ewell RH, Bunting EN, Geller RF. Thermal decomposition of talc. Journal of Research of the National Bureau of Standards. 1935;**15**:551-556

[38] https://en.wikipedia.org/wiki/electrochemical synthesis [Accessed: April 2016]

[39] Rodriguez-Sanchez L, Blanco MC, Lopez-Quintela MA. Electrochemical synthesis of silver nanoparticles. The Journal of Physical Chemistry B. 2000;**104**:9683-9688

[40] Dobre N, Petica A, Buda M, Anicai L, Visan T. Electrochemical synthesis of silver nanoparticles in aqueous electrolytes. UPB Scientific Bulletin, Series B. 2014;**76**(4):1454-2331

[41] Surudzic R, Jovanovic Z, Bibic N, Nikolic B, Miskovic-Stankovic V. Electrochemical synthesis of silver nanoparticles in poly(vinyl alcohol) solution. Journal of the Serbian Chemical Society. 2013;**78**(12):2087-2098

[42] Choi J, Kim K, Han H, Hwang MP, Lee KH. Electrochemical synthesis of red fluorescent silicon nanoparticles. Bulletin of the Korean Chemical Society. 2014;**35**(1):35-38

[43] Khaydarov RA, Khaydarov RR, Gapurova O, Estrin Y, Scheper T. Electrochemical method for the synthesis of silver nanoparticles. The Journal of Nanoparticle Research. 2009;**11**:1193-1200

[44] Sharma S, Choudhary K, Singhal I, Saini R. Synthesis of silver nanoparticles by 'electrochemical route' through pure metallic silver electrodes, and evaluation of their antimicrobial activities. International Journal of Pharmaceutical Sciences and Research. 2014;**28**(2):272-277

[45] Islam AM, Islam MS. Electro-deposition method for platinum nanoparticles synthesis. Engineering International, Asian Business Consortium. 2013;**1**(2):9

[46] http://shodhganga.inflibnet.ac.in/bitstream/10603/21144/10/10_chapter%203.pdf [Accessed: April 2016]

[47] Khah V, Sara D, Djafar IR, Rahman N, Jafar IR. Aglance on the plasma synthesis methodologies of the nanoparticles. In: Proceedings of the Sixth NanoEurope Congress and Exhibition; St. Gallen, Switzeland; 2008

[48] Horikoshi S, Serpone N. Introduction to Nanoparticles, Microwaves in Nanoparticle Synthesis. 1st ed: Wiley-VCH Verlag GmbH & Co. KGaA; 2013

[49] Riaz U, Ashraf SM, Madan A. Effect of microwave irradiation time and temperature on the spectroscopic and morphological properties of nanostructured poly(carbazole)

synthesized within bentonite clay galleries. New Journal of Chemistry. 2014;**38**:4219-4228

[50] Mohammadib M, Imanieha H, Ghammamya S. Rapid and efficient oxidation of organic compounds in microwave condition with new phase transfer oxidative agent: CTAMABC; 2008. DOI: 10.3390/ecsoc-12-01260

[51] Mohan SB, Behera TP, Kumar BVVR. Microwave irradiation versus conventional method: Synthesis of benzimidazolyl chalcone derivatives. International Journal of ChemTech Research. 2010;**2**(3):1634-1637

[52] Joseph S, Mathew B. Synthesis of silver nanoparticles by microwave irradiation and investigation of their catalytic activity. Research Journal of Recent Sciences. 2014;**3**:185-191

[53] Tiwary KP, Choubey SK, Sharma K. Structural and optical properties of ZnS nanoparticles synthesized by microwave irradiation method. Chalcogenide Letters. 2013;**10**(9):319-323

[54] Singh J. Materials Today 2, 2001:10

[55] Okitsu K, Nishimura R. Sonochemical reduction method for controlled synthesis of metal nanoparticles in aqueous solutions. In: Proceedings of 20th International Congress on Acoustics, ICA 2010;23-27. August 2010; Sydney, Australia; 2010

[56] Obreja L, Foca N, Popa MI, Melnig V. Alcoholic reduction platinum nanoparticles, synthesis by sonochemical method, biomaterials in biophysics. Medical Physics and Ecology. 2008:31-36

[57] Rao YN, Banerjee D, Datta A, Das S|K, Guin R, Saha A. Gamma irradiation route to synthesis of highly re-dispersible natural polymer capped silver nanoparticles. Radiation Physics and Chemistry. 2010;**79**:1240-1246

[58] Marignier J, Belloni J, Delcourt M, Chevalier J. New microaggregates of non noble metals and alloys prepared by radiation induced reduction. Nature. 1985;**317**:344-345

[59] Abidi W, Remita H. Gold based nanoparticles generated by radiolytic and photolytic methods. Recent Patents on Engineering. 2010, 2010;**4**(3):170-188

[60] Temgire MK, Bellare J, Joshi SS. Gamma radiolytic formation of alloyed Ag-Pt nanocolloids. Advances in Physical Chemistry. 2011; 9 p. Article ID: 249097

[61] Jayashree B, Ramnani SP, Tewari R. Short aspect ratio gold nanorods prepared using gamma radiation in the presence of cetyltrimethyl ammonium bromide (CTAB) as a directing agent. Radiation Physics and Chemistry. 2010;**79**(4):441-445

[62] Gachard E, Remita H, Khatouri J, Keita B, Nadjo L, Belloni J. Radiation-induced and chemical formation of gold clusters. New Journal of Chemistry. 1998;**22**(11):1257-1265

[63] Krklješ A. Radiolytic synthesis of nanocomposites based on noble metal nanoparticles and natural polymer, and their application as biomaterial (IAEA-RC--12071). International Atomic Energy Agency (IAEA); 2011

[64] Chen Q, Shi J, Zhao R, Shen X. Radiolytic syntheses of nanoparticles and inorganic-polymer hybrid microgels (IAEA-RC--11242). International Atomic Energy Agency (IAEA); 2010

[65] Hornebecq V, Antonietti M, Cardinal T, Treguer-Delapierre M. Stable silver nanoparticles immobilized in mesoporous silica. Chemistry of Materials. 2003;**15**(10):1993-1999

Chapter 3

Physicochemical Aspects of Metal Nanoparticle Preparation

Libor Kvitek, Robert Prucek, Ales Panacek and Jana Soukupova

Abstract

Physicochemical properties, including optical properties or catalytic activity, and biological properties of metal nanoparticles are considerably influenced by their diameter. Therefore, a tailored synthesis of metal nanoparticles represents a key topic in the field of nanotechnology, and the number of research papers, concerning this topic, has been annually growing with an arithmetic progression. Metal nanoparticles are most frequently prepared via chemical reduction of metals in ionic form from their solutions. Using this synthetic approach, tailored parameters of the particles can be achieved via the adjustment of numerous factors: difference of potentials of the metal redox system and the reducing agent redox system, pH of the reaction mixture, and its temperature. The influence of these three factors on the diameter of the prepared metal nanoparticles will be discussed in the following chapter with respect to general laws and based on numerous examples from research practice.

Keywords: metal nanoparticles, tailored preparation, size distribution, chemical reduction, redox potential, pH, temperature

1. Introduction

Metal nanoparticles can be classified among the most studied nanomaterials due to their numerous potential applications [1–3]. Silver and gold nanoparticles have found their targeted applications in the enhancement of Raman scattering due to the optical properties that are associated with the existence of localized surface plasmon resonance (LSPR) [4–8] with the absorption maximum in visible part of the electromagnetic part of the spectra. Thanks to this fact the particles provide a significant enhancement of the Raman signal used in the highly sensitive analytical method of surface-enhanced Raman spectroscopy (SERS) [9–12] used in biology and medicine [13–17]. Transitional metals are commonly known for their high catalytic activity, which is even amplified by the nanodimension of the metal nanoparticles with high ratio between the surface area and the volume of the particle because the catalytic process is located on the surface [18–20]. From the application point of view, even the magnetic behavior of the metal nanoparticles must be taken into account [21, 22]. Last, but not least, the biological activity of the metal nanoparticles must be mentioned, especially in the case of the silver nanoparticles [23–25]. These particles became one of the phenomena of nanotechnology. Recently, more and more products of everyday usage involve these particles, which are applied with respect to their high antibacterial activity against most of the pathogenic bacteria, fungi, and candida [26–28]. As the metal nanoparticles represent a material, which is of high importance, from the research and application point of view, this chapter will be devoted to the methods of their tailored preparation via wet chemical reduction methods. The attention will be paid to the influence of the physicochemical conditions of the particular chemical reaction, to the influence of redox potentials and pH, and to the influence of temperature on the reduction process.

Nanoparticles are generally defined as 3D objects, where at least one of the dimensions is in the range from 1 to 100 nm [29], which is reflected in an abrupt change of the properties and behavior of the materials in the macroscopic range of diameter. Silver nanoparticles represent a typical example of such a material, which biological activity (i.e., antibacterial activity and toxicity against higher organisms) is significantly increased with the decreasing particle diameter [30–32]. Similarly, optical properties of the metal nanoparticles also significantly depend on the diameter range of the particles. The position of the absorption maximum is shifted to the red part of the spectrum with the growing silver or gold nanoparticle size (**Figure 1a**) [33–36].

The catalytic activity of metal nanoparticles is dependent on their diameter because the smaller the particles are, the greater the specific surface area (SSA) is, and the SSA represents the key parameter in heterogeneous catalysis [37–42]. Aside from all of these dependencies, size dependency is monitored also for magnetic properties [43, 44], electrochemical behavior based on dependency of Fermi level on the nanoparticle diameter [45, 46], and also for physical properties, which are considered constant for macroscopic objects—e.g., melting point [47–49] or heat capacity [50–52]. However, dramatic change in the nanomaterial properties happens when the diameter is below 3 nm (note: some of the authors state 5 nm or even 10 nm), which corresponds to approx. 1000 silver atoms [53–55]. Concerning such tiny particles, the organization of valence electrons is changed, and there is a shift from a band structure typical for metals to the organization HOMO-LUMO typical for molecular mass organization. However, a complete shift from one layout structure to the other one happens at the dimension below 1 nm when the particles contain maximally tens of atoms, and their behavior is strongly influenced by quantum effects, which is not commonly observed in the case of nanomaterials [5, 46, 55–57]. These particles are usually named as clusters, and their abnormal optical properties (i.e., fluorescence) are nowadays in the center of the interest of the studies focused on optical properties of nanomaterials [58–63]. Such tiny particles

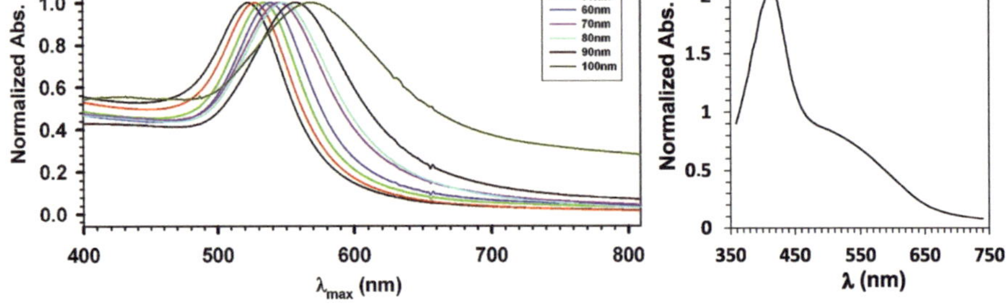

Figure 1.
(a) Dependence of the absorbance of the localized surface plasmon resonance (LSPR) of gold nanoparticles on their size. Reproduced with permission from [33]. (b) Broadening of the LSPR absorbance peak due to the aggregation of 30 nm-sized silver nanoparticles.

represent a unique part of the nanotechnology field, and they will not be discussed any further in the follow-up parts of this chapter.

Aside from the preparation of the nanomaterials, characterization of such a material represents an issue, which will be discussed in considerable detail. Fundamental information about the nanomaterials can be obtained via the measurement of diameter of the particles and their size distribution (also called polydispersity). The characterization methods are either direct or indirect. The direct methods are almost exclusively represented by microscopic methods—electron microscopy and scanning probe microscopy [64, 65]. Intensive development of electron microscopy in the last decade enabled to spread this technique and price availability of scanning electron microscopes (SEM). This characterization method has a limitation of its resolution, which is approx. 10 nm. It disables characterization of smaller particles and does not offer the size distributions of groups of tiny particles. However, one of the greatest advantages of this technique is an easy implementation of energy dispersive spectroscopy (EDS) method [66], which enables qualitative evaluation of chemical composition of the particles. Significantly greater resolution (usually approx. 0.1 nm) can be achieved with the technique of transmission electron microscopy (TEM) [67]. Recently, another technique has become more popular—scanning transmission electron microscopy (STEM). It is a technique, which combines advantages of both electron microscopy techniques, and this way brings new viewpoints into the field of nanomaterial characterization [68].

Electron microscopy represents a method, which requires a good preparation of the characterized sample as it is inserted into a chamber under a high vacuum. It exhibits a problem for metal nanoparticles as they have a strong tendency to aggregation. SEM, with a suitable module even for the metal nanoparticles (i.e., environmental SEM), enables to characterize even wet samples. Similar modules for wet samples are also available for TEM. However, both of the modifications represent economy and technology demanding adjustments, which disables their widespread usage. Another problem with the characterization of metal samples dwells in high electron adsorption by the metal nanoparticles which evokes heating up of the sample and its possible thermal transformation or destruction of the carrying layer of the microscopy grid (carbon, Formvar, etc.). The sample can also involve non-reduced metal ions, which are due to the stream of electrons immediately reduced, and then the observed objects are not truly corresponding with the state of the sample deposited onto the microscopy grid. For example, it is impossible to use electron microscopy for characterization of silver halogenide nanoparticles. The above-listed drawbacks cannot be commonly connected with the second microscopy technique—with the microscopy with scanning probe. The most common adjustment of such a microscopy is in the adjustment of atomic force microscopy (AFM), and it does not require any sophisticated pretreatment of the sample because the characterization does not require vacuum [69]. Due to the specificity of this method (mechanical motion of the scanning probe in a close connection with the surface), this method is more suitable for a study of layers of nanoparticles. If a resolution comparable with TEM is required, the time for gathering sufficiently great amount of data is significantly prolongated [70]. All of the electron microscopy methods are time demanding and require an experienced operator, and therefore they do not prevail over the indirect methods of the nanoparticle characterization. In the case of the metals from the I. B group from the periodic system (Cu, Ag, Au—coinage metals), the nanoscopic characteristics of the prepared particles can be confirmed using a simply qualitative indirect method of UV–vis spectroscopy. Thanks to the size-dependent position and shape of the absorption maximum of LSPR in the visible part of the spectra, it is relatively easy to evaluate qualitatively the size of the prepared nanoparticles and their polydispersity. A typical dependency of the absorption maximum of LSPR on the diameter of spherical particles can be seen in **Figure 1a** [33]. The evaluation of polydispersity of the system can be obtained from the width of the absorption band—the wider the band is, the more polydisperse the system is. UV–vis spectroscopy is such a sensitive technique that there can be observed even more separated absorption maxima corresponding with a highly polydisperse system involving more diameter fractions (multimodal system) [71]. Often it can happen in the case of the aggregated particles, when the larger aggregates can be seen in the spectrum as a flat maximum, which is significantly shifted into the red part of the spectrum compared with the original maximum corresponding with the LSPR of the non-aggregated particles (**Figure 1b**) [72, 73]. The spectrum can be complicated due to the presence of nonspherical particles—e.g., particles with plate morphology will have three absorption maxima in the spectrum, each of them corresponding with the particular diameter of the plate [74].

Quantitative information concerning particle diameter and polydispersity can be obtained from dynamic light scattering method (DLS). This indirect characterization method is based on the diameter determination based on the time-dependent change of the light scattered intensity (using coherent light source) in the system involving particles moving under Brownian motion in the dispersion [75–77]. This method enables quick evaluation of the particle diameter and polydispersity of the system based on enormous number of particles (note: the measurement is realized from tens of seconds up to minutes). This method works in a great diameter range—from 0.3 nm to 3 μm. Unfortunately, DLS is an indirect characterization method based on model behavior of spherical particles, which represent a serious disadvantage for measuring nonspherical particles. Another complication represents a polydisperse system, where the resulting average diameter (Z-average d_Z) is highly influenced by the largest particles present in the system [78, 79]. The problem of differently weighted particles based on their different diameters is neglected in numerous papers, and the difference between the diameter obtained from electron microscopy and DLS is explained as the difference between the real and hydrodynamic diameter. The influence of nanoparticle solvation (or the presence of adsorbed molecules including stabilizers of macromolecular character) on the resulting particle diameter can be in couple of percents, but the presence of large particles in the system is responsible for the difference in tens of percents. In the case of the electron microscopy, all the particles are considered with the same weight, while in the case of DLS, the weight of each particle in the average d_Z is given by the power of five of its diameter [80–83]. The instruments working on the DLS principle are equipped with software, which enables different modes according to which the particles can be weighted (incl. the weighting identical with electron microscopy), but it is necessary to keep in mind that the recalculation is based again on the approximation model and its accuracy, which unfortunately decreases with growing polydispersity of the system [84]. The DLS method is therefore more suitable for monitoring of the processes in the dispersed systems (e.g., processes of particle aggregation) than for the determination of absolute values of diameters of the particles in the system. On the other hand, DLS represents one of the fundamental methods defined in ISO 22412:2017 for the determination of diameter of colloidal particles. From the other commonly used indirect methods of determination of the particle diameter, it is necessary to mention Scherrer method of determination of crystal domains using powder RTG diffraction (XRD). The calculation is based on the dependency of line broadening at half the maximum intensity (FWHM) on the diameter of the nanoparticles [85]. This method does not evaluate a full particle diameter. It determines the size of coherent domains, which the resulting particle is formed of. It is a method, which is highly suitable for the study of aggregates, where the independent particles (grains) joined one another into a large aggregated object.

Metal nanoparticles can be prepared using both physical and chemical methods. Physical methods of the nanoparticle preparations are based on the use of macroscopic objects, which are disintegrated onto the nanoobject using different kinds of mechanical milling procedures, dispergators, or sonicators. Generally, these methods are labeled as "top-down methods." Unfortunately, these methods are not suitable for metal nanoparticle preparation due to the typical mechanical properties of the material (malleability, ductility). Therefore, different kinds of energy, such as the application of laser beam on macroscopic material, are preferred for the preparation of metal nanoparticles. The method is called a laser ablation [86–89]. The nanoparticles are formed due to a local overheating of the metal material, its evaporation, and consequential condensation of the vapors. This process does not need to be initiated by any other substances and as such is able to produce highly clean nanoparticles. The diameter of the generated particles can be tailored due to the energy applied, wavelength, and the laser pulses [90–93]. The diameter of the generated particles and polydispersity of the system can be modified by further application of laser [94, 95]. As the generated dispersions are usually highly diluted, it is not necessary to add any stabilizers into the system. However, stabilizers are usually added into the system as they can prolong temporal stability of the dispersed system, and their presence in the system from the very beginning can significantly influence the particle characteristics [96–98]. The application of laser ablation in practice is limited as the number of produced particles is relatively low; their production is barely enough for research purposes. Comparable transition of macroscopic material to vapors and consequential condensation can be also realized in tube furnace, where the energy flow is directly in a form of heat [99, 100]. The disadvantage of this method is related to possible contamination of the furnace. Moreover, this process is highly energy demanding. Recently, a new method, ranged into the top-down methods, has been introduced. It is a method of vacuum sputtering. This method is based on the bombardment of a target with energetic gas ions generated via the collision of electrons and carrier gas in vacuum using direct current (DC), radio frequency (RF), or magnetron sputtering [101–104]. This method is primarily designed for the generation of nanoparticle layers on the surface of a solid substrate. However, a modification of the process, when the solid substrate is replaced by the surface of a liquid, enables preparation of liquid dispersion of nanoparticles [105–107]. The enumeration of physical methods of nanomaterial preparation cannot be completed without the lithographical techniques. These techniques enable only fabrication of nanostructural layers [108].

Chemical methods of nanoparticle preparations, compared to the physical ones, are based on the reduction of analytical solutions containing the corresponding metal ions to atoms using the appropriate reducing agent. The first neutral atoms form nuclei of the emerging nanoparticles which can quickly grow due to the proceeding reduction of the ions until the diameter of the nanoparticles is reached—LaMer mechanism of the nanoparticle growth [109]. The reduction of the metal ions can be achieved using a variety of reducing agents. Aside from suitable reducing agents, the reduction can be also done using photochemical, radiochemical, or electrochemical approaches. Photochemical approach is usually based on the application of UV or vis irradiation, which generates in the solution, at the presence of suitable organic sensibilizers, radicals that fulfill the role of a reducing agent and reduce the metal ions to neutral metal atoms in the solution [110, 111]. Direct photochemical reduction is possible for some photosensitive materials (e.g., silver halogenides). Free electrons are generated by absorption of photons by these substances, and therefore direct reduction of the metal ions from crystal lattice can proceed in this way [112]. However, in this special case, it is necessary to highlight that the ions are not reduced directly from the solution because the photosensitive compounds have limited solubility. Photoreduction therefore proceeds in solid state [113]. Radiochemical reduction uses highly energetic gamma irradiation, X-ray irradiation, or accelerated electrons and does not need additional organic molecules, which would produce radicals. These kinds of irradiations enable to produce radicals from the solvent molecules [114–116]; eventually the applied accelerated electrons can be responsible for the reduction of the metal ions [117, 118]. Therefore the generated particles are really clean although it is highly recommended to add suitable stabilizers into the systems in order to prevent unwanted aggregation of the primarily generated particles. The reduction of metal ions in solutions can be also achieved using electrochemical methods. In this case, the electrons, needed for completion of the reduction, are supplied by the electrode. The surface of the electrode is then covered by a film consisting of the generated particles [119–121]. One of the biggest advantages of this method is its variability thanks to the possibility of the reducing potential adjustment or possibility to adjust the passing current [122–124]. This method of nanoparticle generation is commonly used for layers of nanoparticles generated on glass with conductive layer of indium tin oxide (ITO) glass [125]. In the case of the liquid dispersion of nanoparticles, prepared via the electrochemical approach, it is necessary to prevent firm attachment of the generated particles on the surface of the electrode. For this purpose the combination of electrochemical reduction with ultrasound is generally used [126]. However, the commonly used approach to electrochemical preparation of nanoparticles is based on the use of a mediator molecule, which is reduced on the cathode, and then the reduced form of the mediator diffuses into the solution where it is used for reduction of the metal ion. Methylviologen is the most commonly used mediator [127–130].

2. Preparation of metal nanoparticles by chemical reduction: role of the redox potentials

Preparation of nanoparticles of transition metals via wet route chemical reduction method is probably the most frequently used one. It is widely used both in research and in technological practice [131–133]. The reasons why this kind of preparation method is used are simple—from the economic and technological point of view, it represents the most convenient approach. The preparation of nanomaterials following the wet chemical reduction method does not have any specific demands for complicated technological equipment or installation, in most of the case just a suitable vessel equipped with stirring mechanism and a possible adjustment of temperature (note: for elevated temperature, heating is needed; for lower than room temperature, simple cooling bath is usually sufficient) is needed. Moreover, the productivity of chemical reduction methods is significantly higher than other methods of metal nanoparticle preparations. The whole synthetic process is easy to scale up from a preparation of a couple of milliliters of the reaction mixture up to tens of liters. Another advantage is also a wide range of available reducing agents. The reducing agents are commonly inexpensive substances, and therefore the economic balance of the whole process is not anyhow weighted down in this aspect. Also the energy balance of the whole process is in favor of the wet chemical reduction method, compared to the physical methods of metal nanoparticle preparation. However, a certain disadvantage can be found in the limited possibility to design the particle diameter. This handicap has been almost erased thanks to the intensive research of this aspect in the last 20 years. Additionally, the research brought also a possibility to adjust other fundamental properties like morphology, which is highly problematic to achieve even via physical approach to metal nanoparticle synthesis [134, 135].

From the thermodynamic point of view, the synthesis of metal nanoparticles via wet chemical reduction method can spontaneously proceed only under the condition of negative value of the difference of Gibbs energy (ΔG). To achieve such negative ΔG values, the difference of redox (reduction-oxidative) potentials of the redox systems metal ions and oxidized and reduced form of the reducing substance must be positive. As the primary process of the electron transfer between the redox forms of the redox system is reversible in its principle, the description of the proceeding processes can be done thanks to the analogy to the processes proceeding on the electrodes in a galvanic cell. The difference in Gibbs energy (ΔG) for the reaction in a galvanic cell is given by the difference of equilibrium potentials of the electrodes (ΔE) according to Eq. (1):

$$\Delta G = -\, zF\, \Delta E \quad (1)$$

where *F* is Faraday constant and *z* represents the number of electrons exchanged among the reactants in cell reaction [136]. The ΔG has to be negative for spontaneous course of the reaction, and therefore the ΔE—i.e., the difference of the potentials of the reacting redox systems—has to be positive. According to the convention, this difference of the electrode potentials is defined as the difference between the right and the left electrode, whereas the oxidation proceeds on the left and the reduction on the right electrode. If the metal ion reduces and the molecule of the reducing agent is being oxidized, the ΔE can be defined by Eq. (2):

$$\Delta E = E_R - E_L = E_{Mez+/Me} - E_{ox/red} \quad (2)$$

The potentials of both redox systems $E_{Mez+/Me}$ and $E_{ox/red}$ depend on the ratio of the activity (concentration) of the oxidized and reduced form according to the Nernst equation in Eq. (3):

$$E_{ox/red} = E^{o}_{ox/red} + (RT/zF)\ln(a_{ox}/a_{red}) \quad (3)$$

where $E^{o}_{ox/red}$ represents the value of standard redox potential for the particular redox system in the hydrogen scale (standard redox potential of the hydrogen electrode was determined to the value of 0) and a_{ox} and a_{red} represent activities of the oxidized and the reduced forms of the given redox system in the solution. The activity of a solid metal is unitary, and although the real value of $E^{o}_{Mez+/Me}$ is dependent on the particle diameter for little nanoparticles [137], this dependency would not be taken into account in these considerations. More complicated situation is the estimation of the redox potential of the reducing agent. Although the primary process of electron transfer can be considered reversible, the oxidized form of the molecule of the reducing agent is usually unstable and undergoes consequential chemical reactions with the substances available in the solution. Therefore, the process becomes irreversible, and it is almost impossible to determine the concentration of the oxidized form. Thus, the fundamental decision if the redox process can be realized is based on the value of ΔE^{o} instead of ΔE. It is relatively a rational consideration because if the ΔE^{o} has a negative value, then positive value of ΔE of both redox systems cannot be achieved. Such a large difference would be necessary for a quick course of the redox procedure [138]. The values of redox potentials of the discussed metal systems are listed in **Table 1**.

It is obvious that noble metals like silver, gold, palladium, or platinum can be reduced from their salts using even relatively weak reducing agents. However, for the reduction of less noble metals like copper and especially iron, cobalt, and nickel, only stronger reducing agents, e.g., sodium borohydride, must be used. It is

Redox system	Half-reaction	E° [V]	Redox system	Half-reaction	E° [V]
Iron	$Fe^{2+} + 2e^- \rightleftharpoons Fe^o$	−0.447	Ruthenium	$Ru^{2+} + 2e^- \rightleftharpoons Ru^o$	+0.455
Cobalt	$Co^{2+} + 2e^- \rightleftharpoons Co^o$	−0.28	Rhodium	$Rh^{3+} + 3e^- \rightleftharpoons Rh^o$	+0.758
Nickel	$Ni^{2+} + 2e^- \rightleftharpoons Ni^o$	−0.257	Palladium	$Pd^{2+} + 2e^- \rightleftharpoons Pd^o$	+0.951
Copper	$Cu^{2+} + 2e^- \rightleftharpoons Cu^o$	+0.3419	Osmium	$Os^{2+} + 2e^- \rightleftharpoons Os^o$ Ref. [140]	+0.7
Silver	$Ag^+ + e^- \rightleftharpoons Ag^o$	+0.7996	Iridium	$Ir^{3+} + 3e^- \rightleftharpoons Ir^o$	+1.156
Gold	$Au^{3+} + 3e^- \rightleftharpoons Au^o$	+1.498	Platinum	$Pt^{2+} + 2e^- \rightleftharpoons Pt^o$	+1.18

Table 1.
Standard electrochemical potentials of noble metals and triad Fe, Co, and Ni [139]. All potentials are referred vs. standard hydrogen electrode (SHE) using IUPAC recommendation.

Redox system	Half-reaction	E° [V]	Cit.
Hypophosphorous acid	$HPO_3^{2-} + 2H_2O + 2e^- \rightleftharpoons H_2PO_2^- + 3OH^-$	−1.65	[139]
Borohydride	$H_2BO_3^- + 5H_2O + 8e^- \rightleftharpoons BH_4^- + 8OH^-$	−1.24	[139]
Hydrazine	$N_2 + 4H_2O + 4e^- \rightleftharpoons N_2H_4 + 4OH^-$	−1.15	[140]
Hydroxylamine	$N_2O_2^- + 6H_2O + 4e^- \rightleftharpoons 2NH_2OH + 6OH^-$	−0.73	[140]
Formic acid	$CO_2 + 2H^+ + 2e^- \rightleftharpoons HCOOH$	−0.14	[140]
Hydrogen peroxide	$O_2 + H_2O + 2e^- \rightleftharpoons HO_2^- + OH^-$	−0.076	[140]
Ascorbic acid	$C_6H_6O_6 + 2H^+ + 2e^- \rightleftharpoons C_6H_8O_6$	−0.054 (pH = 7)	[140]
Formaldehyde	$HCOOH + 2H^+ + 2e^- \rightleftharpoons HCHO + H_2O$	−0.02	[140]
Hydrogen	$2H^+ + 2e^- \rightleftharpoons H_2$	0.000	[140]
Stannum(II)	$Sn^{4+} + 2e^- \rightleftharpoons Sn^{2+}$	+0.15	[140]
p-Aminophenol	$OC_6H_4NH + 2H^+ + 2e^- \rightleftharpoons HOC_6H_4NH_2$	+0.599	[149]
Hydroquinone	$C_6H_4O_2 + 2H^+ + 2e^- \rightleftharpoons C_6H_4(OH)_2$	+0.699	[140]
Citric acid	$CH_3COCH_3 + 3CO_2 + 2H^+ + 2e^- \rightleftharpoons C_6H_8O_7$	+1.1	[150]
Glucose	$C_6H_{12}O_7 + H_2O + 2e^- \rightleftharpoons C_6H_{12}O_6 + 2OH^-$	≈ + 0.65 (pH = 12.8)	[151]
Galactose	$C_6H_{12}O_7 + H_2O + 2e^- \rightleftharpoons C_6H_{12}O_6 + 2OH^-$	≈ + 0.7 (pH = 12.8)	[151]

Table 2.
Standard electrochemical potentials of various reductants commonly used in the synthesis of metal nanoparticles. All potentials are referred vs. standard hydrogen electrode (SHE) using IUPAC recommendation.

not realistic to think that the nanoparticles of less noble metals can be prepared via green reducing methods using reducing substances of biological origin [141–147], in this case, substances of a type of polyphenols, the redox potential is insufficient with respect to the highly negative values of standard redox potentials of the metal systems [148]. The values of redox potentials of the commonly used reduction agents are listed in **Table 2**.

Figure 2 shows graphically comparison of redox potentials of some reducing agents and metals. The reducing agent can be effective, in the process of the metal ion reduction, only if the redox potential is under the redox potential of the metal

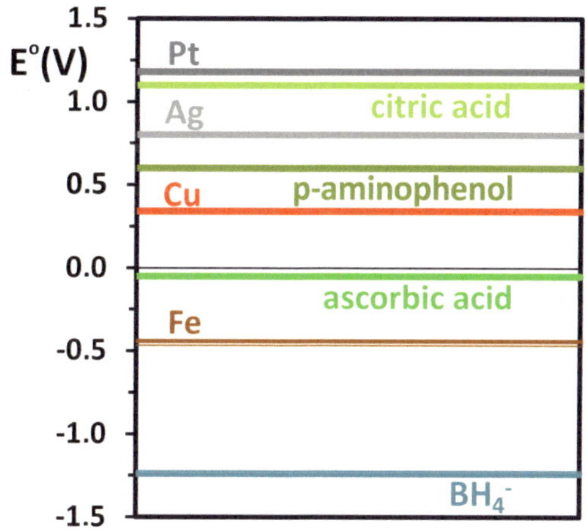

Figure 2.
Comparison of the standard redox potentials of the metals and reductants selected from ***Tables 1*** *and* ***2*** *as representative of the typical reaction systems for the preparation of the metal nanoparticles.*

redox system on the hydrogen scale. However, any positive value of ΔE is not sufficient for real and effective course of the reduction leading to the formation of metal nanoparticles under room temperature. The value of ΔE from approx. 0.15–0.3 V is commonly agreed to ensure sufficiently quick course of the reaction. Based on all of the stated facts, it can be concluded that the impact of the ΔE value is crucial. The higher the value of ΔE is (i.e., the stronger the reducing substance is), the quicker the reduction of the metal ions is, and the generated nanoparticles are smaller [152]. Due to a quick reduction, steep growth of oversaturation proceeds. According to the theory of new phase formation, a great number of nucleation centers are generated under such conditions. The nuclei grow quickly, thanks to the aggregation mechanism in majority, up to the final particle diameter within a couple of seconds. The result of such a process is a formation of tiny nanoparticles, which is a typical example of the use of sodium borohydride as the reducing agent, which enables the synthesis of silver nanoparticles, gold nanoparticles, and nanoparticles of other metals that have the diameter smaller than 10 nm. The first usage of sodium borohydride, as the reducing agent for the synthesis of silver nanoparticles, was described by Creighton, Blatchford, and Albrecht in the 1980s. This collective of authors used sodium borohydride for the preparation of SERS substrate [153]. Unfortunately, a quick reduction of available metal ions has negative consequences such as poor reproducibility of the synthesis and high polydispersity of the generated particles in the system. Therefore, the synthesis is likely to proceed under lowered temperature (note: the system is placed into a cooling bath) and with the addition of stabilizing substances. In the borohydride reduced systems, citrate molecules are frequently used as stabilizing agent as it prevents aggregation of the generated particles. The citrate ions can be, in the next step, easily removed from the surface layers due to the addition of different molecules in the process of nanoparticle functionalization [154–156]. Strong reducing agents cannot, however, guarantee synthesis of nanoparticle systems with minimal polydispersity of the generated particles as the two phases of the nanoparticle formation—nucleation and particle growth—are not well differentiated. Not even the usage of weak reducing agents, requiring elevated temperature for its effective function, can solve this problem. The process of silver nanoparticle preparation introduced by Lee and Meisel is based on the reduction of silver ions with citrate, boiling the reaction mixture [157]. Thanks to high kinetic energy (boiling water), new nuclei can be generated even in further stages of the reduction process, and the resulting system is then highly polydisperse.

The average diameter of the generated particles is due to the slower course of the reaction, significantly bigger (40–50 nm) than for the borohydride method [158, 159]. However, when the citrate anion is used as the reducing agent of Au^{3+} cations (Turkevich method [160–162]), the greater difference in redox potentials of both of the systems predetermines significantly quicker reduction of the particles and generation of smaller particles (approx. 15 nm) with even better polydispersity of the system. Therefore, it is necessary to compare the results achieved with different reducing agents only for one metal redox system where the growing diameter of the generated particles can be unambiguously put into context with the decreasing reducing strength of the reducing agent. This behavior was observed also for the silver nanoparticles prepared via the modified Tollens process using different reducing saccharides with different reducing ability [30, 163, 164].

The above-listed examples proved that the course of the metal ion reduction, leading to the production of metal nanoparticles, is given by the ΔE more than the value of $E_{red/ox}$ of the used reducing agent. However, the estimation of the ΔE value (note: accurate calculation is rather complicated or even impossible) is complicated by the dependency of redox potential of the reducing agent and very frequently also of the metal redox system on pH, which will be discussed in the independent part. It is also necessary to take into account the interaction of metal ions with other substances available in the reaction system as they can form complex substances with these ions. Such substances—ligands—can be added into the reaction system on purpose. However, sometimes they can be also counterparts of a complicated, not well-defined biological system of the reducing system (in the case of the green methods) [165–167]. In other case, the reducing substance can play the role of a reducing agent and a ligand, which forms a complex compound with the available metal ion. It is typical for reducing polymers including a nitrogen atom as heteroatom, e.g., poly(vinylpyrrolidone) (PVP), polyethylenimine (PEI), or polyaniline (PANI) [168–174]. Such a formation of a complex compound is widely applied in practice for the tailored preparation of metal nanoparticles. Complexing agent can influence reduction rate of the metal ion due to the change of redox potentials via complexation of metal ions (see some examples of potential change on **Figure 3**). This way it significantly influences the nuclei formation and their consequential growth. The synthesis of silver nanoparticles, using the reduction of diammonium silver complex cation with reducing sugars (already mentioned modified Tollens

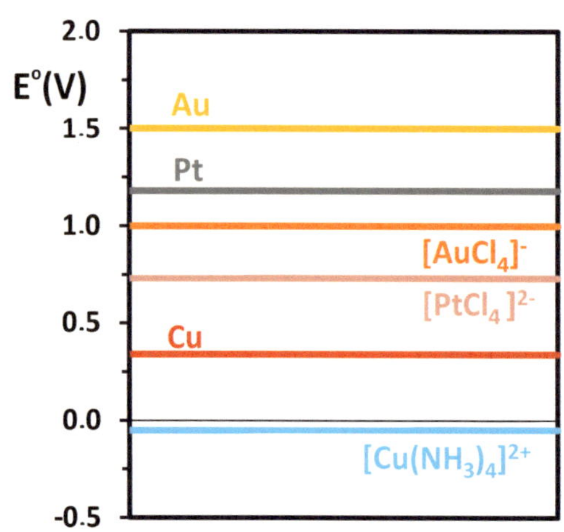

Figure 3.
Comparison of the standard redox potentials of the free metal ions and their typical complex ions, which are employed in the reaction systems for the preparation of the metal nanoparticles. Values of standard redox potentials of the complex ions are taken from tables by Dobos [140].

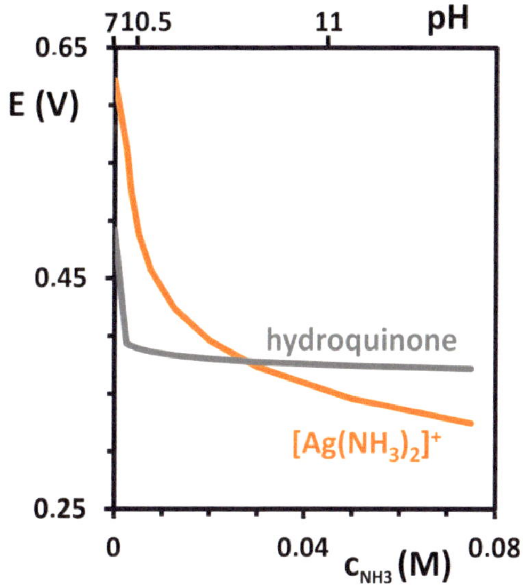

Figure 4.
Comparison of the redox potential dependence for the complex metal ion $[Ag(NH_3)_2]^+$ and hydroquinone reductant on the concentration of ammonia. The change of the ammonia concentration in the reaction system is connected with the change of pH. It is clearly visible that at pH ≈ 10.8 (concentration of ammonia about 0.0275 Mol/dm^3), the potentials of both redox systems are identical. Values of redox potentials were calculated using Eq. (10) for system Q/H_2Q and Eq. (4) for the system $[Ag(NH_3)_2]^+/Ag$. The needed values of constants were taken from tables by Dobos [140].

reaction), represents a suitable example of such a tailored metal nanoparticle preparation. The diameter of the generated silver nanoparticles can be adjusted by means of the ammonia concentration throughout the change of the redox potential of the silver redox system due to the bond of majority of the Ag^+ cations into a stable complex compound $[Ag(NH_3)_2]^+$ according to Eq. (4):

$$E_{[Ag(NH3)2]+/Ag} = E^o_{Ag+/Ag} + (RT/F)\ \ln\{(a_{[Ag(NH3)2]+})(a_{NH3})^{-2}\beta_2^{-1}\} \quad (4)$$

where β_2 represents a stability constant of the complex cation $[Ag(NH_3)_2]^+$ [175]. **Figure 4** shows how is such a difference reflected in the course of the potential of silver redox system side by side with the course of the difference of the potential of the used reducing agent with the changing pH value (with the growing ammonia concentration in the reaction system). The growing ligand concentration in the reaction system is reflected in a decrease of ΔE, decrease in the rate of the reaction, and growth of the particle diameter [163, 176–179]. Also ligand exchange for any other, which would form a stronger complex with the metal ion, will lead to a decrease of ΔE and consequential decrease of the reaction rate, which will lead to the growth in the diameter of the finally generated particles. This behavior was observed in the case of silver nanoparticles, prepared with the assistance of sulfite ligand as the complexing agent. This ligand has six order higher stability constants than the values for ammonia complex with this metal [175]. Therefore, the growth of the size of the prepared particles was more than significant. The generated particles had several hundreds of nanometers in diameter [180].

3. Preparation of metal nanoparticles by chemical reduction: role of the pH

The tailored preparation of metal nanoparticles using the method of chemical reduction not be done without the control of pH of the system. The adjustment of pH helps to manage the diameter of the prepared particles and reproducibility of the reducing process. The influence of pH is connected with the fundamental principles:

1. pH usually influences redox potential of the used reducing agent and can significantly change even the mechanism of its oxidation.

2. pH influences redox potential of the metal, which is usually based on the formation of a new compound, where the reduced metal ion is bound.

3. pH significantly influences stability of the generated metal nanoparticle dispersion by means of the change of their zeta potential, or via the electric properties of the stabilizing molecules, e.g., via a change in the dissociation of the function groups in the polymer electrolytes.

The first two mentioned principles are the most important. As the diameter of the generated particles can be tailored due to the strength of the reducing agent, this aspect will be devoted more attention. The simplest description of the change of redox potential with pH is for the system where gaseous hydrogen is used as the reducing agent. If the solution, containing ionic metal, is saturated with gaseous hydrogen under the atmospheric pressure, the equation for the redox potential H^+/H_2 is significantly simplified:

$$E_{H+/H2} = (RT/F)\ \ln\ (a_{H30+}) = -(2303RT/F)\ pH \qquad (5)$$

If pH increases from deep acidic value 1 to neutral value of 7, the original redox potential is shifted from 0.0 to −0.414 V. As the reduction potential of hydrogen is significantly increased, the reduction will proceed quicker, and the average particle diameter will be smaller [181].

In the case of one of the strongest inorganic reducing substances, used for the metal nanoparticle generation, the change of pH from acidic region to alkali one does not mean just the change in redox potential but also a change in the oxidation mechanism of this substance. The oxidation of the borohydride anion proceeds as follows in the alkali environment [139]:

$$BH_4^- + 8OH^- \rightleftharpoons BO_2^- + 6H_2O + 8e^- \quad E^o = -1.24\ V \qquad (6)$$

However, if the pH is shifted into a slightly acidic region, the reaction mechanism is significantly changed, and the oxidation of the borohydride anion can be simply written as follows:

$$BH_4^- + 3H_2O \rightleftharpoons B(OH)_3 + 7H^+ + 8e^- \quad E^o = -0.48\ V \qquad (7)$$

Additionally, the environment enables also a side reaction—direct hydrolysis of the borohydride anion accompanied by the hydrogen release:

$$BH_4^- + 2H_2O \rightarrow BO_2^- + 4H_2 \qquad (8)$$

In this situation, hydrogen, generated according to Eq. (8), can become the main reducing agent. However, such a complicated reaction of the tailored nanoparticle synthesis and the reproducibility of such synthetic experiment are rather poor. The dependence of oxidation mechanism and redox potential on pH can be observed for the system of hydrazine and hydroxylamine [182]. Both of these substances are commonly used as the reducing agents in the processes of nanoparticle preparation [183–189].

Organic reducing substances, used in the preparation of metal nanoparticles, exhibit more predictable behavior than the inorganic ones. These substances are weak electrolytes, in majority, and as such their dissociation is influenced by pH. Generally, the substances, which behave as weak acids, have stronger reducing ability in alkali region of pH, and those that behave like weak bases have stronger reducing ability in acidic pH range. It is connected with higher degree of dissociation of these reducing substances in the particular pH region. In literature, *p*-dihydroxybenzene (also labeled as hydroquinone or H_2Q) is usually stated as a typical representative of such a substance [190]. Its dissociated form is oxidized, in two one-electron steps, across semiquinone, to *p*-benzoquinone (often labeled as quinone or Q):

$$C_6H_4O_2 + 2H^+ + 2e^- \rightleftharpoons C_6H_4(OH)_2 \text{ resp. } Q + 2H^+ + 2e^- \rightleftharpoons H_2Q \tag{9}$$

As H_2Q is a weak acid ($pK_1 = 9.85$ a $pK_2 = 11.4$), the concentration (resp. activity) of the Q^{2-} ion is dependent on pH, which is then reflected in the equation for potential of the Q/Q^{2-} redox system:

$$\begin{aligned} E &= E^{o}_{Q/Q2-} + (RT/2F)\ln(a_Q/a_{Q2-}) \\ &= E^{o'}_{Q/Q2-} + (RT/2F)\ln([H^+]^2 + K_1[H^+] + K_1K_2) \end{aligned} \tag{10}$$

where $E^{o'}_{Q/Q2-}$ is a formal redox potential of the Q/Q^{2-} redox system and K_1 and K_2 are partial dissociation constants of H_2Q [139]. Quinone is not an electrolyte, and therefore its dissociation is not a part of the equation of the influence of redox potential on pH. In the case of other organic substances, both redox forms can be electrolytes, and therefore the equation of the dependency of redox potential on pH is significantly more complicated as both of the dissociations must be involved. However, the real mechanism of the H_2Q oxidation is more complicated due to the formation of semiquinone, but the pH dependence of the redox potential is not influenced [191–193]. The dependency of the redox system of Q/Q^{2-} on pH side by side with the dependency of the potential of the redox system Ag^+/Ag on the ammonia concentration is shown in **Figure 4**. It is obvious that the pH and the concentration of the complexing agent for ions influence ΔE of the reaction, its reaction rate, and also the diameter of the generated metal nanoparticles. In this case, when H_2Q was used as the reducing agent for Ag^+ ions, resp. $[Ag(NH_3)_2]^+$ at pH lower than 7, the value of ΔE is low and the reduction of silver ions does not proceed [194, 195]. Similar dependency of redox potential on pH as H_2Q can be observed also with ascorbic acid (AA). As it is a stronger acid than H_2Q, a certain reducing ability can be observed already at pH lower than 7 (approximately pH = 6), but it does not behave as an effective reducing agent until pH 9 is reached. In such a case, the reduction is completed within several minutes, and the generated silver nanoparticles (AgNPs) have the diameter smaller than 50 nm [11, 196, 197]. Weak reducing agents, e.g., reducing saccharides, are able to reduce silver ions only under pH higher than 7. However, truly defined particles can be generated under much higher pH, i.e., under pH around 12 [30, 163, 198–200]. Also polymers are typical representatives of substances with a weak reducing ability which can be achieved only under specific conditions as, e.g., optimal pH for the reduction [201–208]. Pluronic P123 (a copolymer of PPO and PEO) has a slight reducing ability due to the polyethylene oxide blocks (PEO). The reducing ability grows with the growing pH. This polymer is not able to perform reduction of $AuCl_4^-$ ions in acidic region. However, with the growing pH, it is possible to generate gold nanoparticle dispersion with the average particle diameter of 10 nm and the SPR in the region of approx. 550 nm when the final pH of the dispersion is higher than 10 [209].

In the case of the metal redox systems, the influence of the pH is related to the reaction of the OH^- ions with the metal ions. Hydroxides, oxidehydroxides, oxides, or stable complex substances are commonly formed by this reaction. A typical example represents formation of poorly soluble Ag_2O in Ag^+ solution at pH higher than 9. The redox potential of the system Ag_2O/Ag is significantly lower (E^o = 0.342 V) than of the system Ag^+/Ag (E^o = 0.799 V). This change of redox potential is reflected in a lower rate of reduction and a formation of big particles. In the case when glucose is used as the reducing agent, the generated particles are even larger than 100 nm [181, 194, 195, 210, 211].

On the contrary, in the case of gold chloride, respectively complex anion $AuCl_4^-$, it is possible to exchange coordinated Cl^- ions for OH^- anions in aqueous environment, according to the adjusted pH. The complex anion of the composition of $AuCl_3(OH)^-$ is formed under the pH lower than 6. All the Cl^- anions are replaced by the OH^- anions when the pH becomes higher than 12. The newly formed complex anions are more stable than the original ones, and therefore the reaction rate is significantly reduced due to the decrease of the redox potential of the system Au^{3+}/Au. It is then reflected in the growth of the AuNPs' diameter [212]. In the case of the Turkevich method of AuNPs' preparation, the situation is complicated with the way the pH changed. It proceeds with a growing concentration of citrate, which is added as reducing agent. Therefore, although the newly generated anion $AuCl_{4-x}(OH)_x^{x-}$ is more stable, the reduction rate can grow with the growth of the reducing agent concentration. This factor together with the positive dependence of the reduction power citrate on the pH can prevail, and then the diameter of the generated AuNPs decreases with the growing pH [213]. It must be taken into account that the pH value can influence both redox pairs in return. Therefore, the prediction of the diameter dependence of the prepared nanoparticles on pH represents a highly complicated task in some cases.

4. Preparation of metal nanoparticles by chemical reduction: role of the temperature

The temperature is the last physicochemical parameter, which can have an impact on the course of the metal nanoparticle synthesis and which is therefore worth discussing in this chapter. It is generally known that temperature influences both chemical balance and rate of chemical reaction. The first case cannot be, however, used as the one, which has a major influence on the metal nanoparticle preparation via a reducing process. This fact is caused by the minimal changes of the potential values, i.e., ±1 mV/K [214]. This difference in ΔE, in favor of the products of the reduction process, is too minor to have any significant impact. The change in temperature, however, significantly influences the rate of the chemical reaction. The increase in temperature leads to the increase in the reduction rate according to the Arrhenius equation. As it was already mentioned before, the mechanism of the nanoparticle formation fundamentally consists of two steps: (1) generation of nuclei and (2) growth of the nuclei up to the size of a final nanoparticle [215, 216]. Both of the stages of the nanoparticle formation require different activation energies, and therefore their dependency on temperature differs [212, 217]. The nuclei formation is more energy demanding, and therefore the increase of temperature in this step is reflected in steeper increase of its rate than the rate of the nanoparticle growth. The increase in temperature is then reflected in the increase of the polydispersity of the system because new nuclei will be generated for the whole reaction time side by side with the growing particles. This effect can be especially well observed in the case of the weaker reducing agents as, e.g., for the preparation of silver nanoparticles according to Lee and Meisel [157]. This effect is even more significant concerning very weak reducing agents of polymer character [204, 205]. Contrariwise, the average particle diameter decreases with the increased rate of the nuclei formation [218, 219].

Higher temperature is used especially with weaker reducing agents with the aim to increase the rate of chemical reaction and eventually to increase the salt conversion. On the contrary, the use of stronger reducing agents requires a decrease in this parameter—i.e., cooling of the reaction mixture can positively influence also polydispersity of the generated particles in the system. The improvement of the system polydispersity is predetermined by two contributions: first, by a limited generation of the future nanoparticle nuclei, and second, by a limited Brownian motion of the generated particles. Lowered temperature is used commonly especially in cases when borohydride is used as the reducing agent. The original procedure suggested by Creighton, Blatchford, and Albrecht used ice-cold bath [153]. However, cooling with ice is insufficient for suppression of the quick generation of the nuclei in the course of the whole reaction time. Therefore, neither polydispersity nor reproducibility of the resulting nanoparticles are not satisfactory. Both parameters were further improved when the addition of citrate anion was introducing into this already established synthesis. The citrate does not play the role of reducing agent at this temperature, but it prevents the unwanted aggregation of generated particles. As such, it fulfills the role of a stabilizing agent [156]. To reach lower temperatures than 0°C, it is necessary to use other solvents but just water. The temperature of −25°C can be achieved when water is replaced by mixed solvent—water-ethanol (1:1 v/v). However, the particles generated in this mixed solvent exhibit even worse aggregation stability than the particles prepared in a pure aqueous environment. Therefore, it is necessary to use a stabilizer—in this case, sodium polyacrylate was introduced. Under such a low temperature, the ability of the nuclei formation is significantly limited, which is then reflected in the increased diameter of the generated nanoparticles—3.4 nm for −5°C and 7.2 nm for −25°C. The polydispersity of the system, however, follows the already discussed trend, i.e., it is decreased with the decreasing temperature [220].

5. Conclusion

Metal nanoparticles represent one of the key pillars of nanotechnology. Their physicochemical properties are unique and different to other types of nanomaterials that they have abruptly found their way to numerous commercial applications. However, the applications are strongly dependent on the efficiency of their production—especially respecting primarily defined diameter and morphology. Nowadays available methods of metal nanoparticle production do not fully fulfill the needs given by the research and practice. In order to design preparation of tailored nanoparticles, it is highly important to be aware of the impact of the different physicochemical parameters predetermining the diameter and morphology of the generated nanoparticles. Let us name at least a couple of them—the influence of the difference of redox potentials of the reaction components, pH of the reaction mixture, and its temperature. These three parameters have been discussed in this chapter with respect to the current state of knowledge. However, aside from the three discussed parameters, there are several other parameters with key influence. One is molar ratio among the reactants, which was already effectively used by Turkevich for a tailored preparation of gold nanoparticles [154, 160, 212, 221, 222]. The whole reduction process can be also influenced by other substances present in the reaction system—e.g., molecules of surfactants, polymers, and other low-molecular organic substances with O, N, and S heteroatoms, which significantly influence the process of nucleation and growth of the nanoparticles [223–226] and also the aggregation stability of the final system [227–229]. The process of "seed-mediated" preparation of nanoparticles is based on the preparation of seeds, in diameter of nanometers, using strong reducing agents. The nuclei are, in the next step, let to grow into their required diameter, thanks to the addition of new portion of the metal ions reduced by using weaker reducing agent [11, 230, 231]. Reasonable adjustment of both steps, nanoparticles in a wide range of diameters, can be prepared. Additionally, this approach also provides dispersions of nanoparticles with reasonably good values of polydispersity [31, 232, 233].

Acknowledgements

The work was supported by ERDF project "Development of Pre-applied Research in Nanotechnology and Biotechnology" (No. CZ.02.1.01/0.0/0.0/17_048/0007323).

Conflict of interest

The authors certify that there is no actual or potential conflict of interest in relation to this article.

Author details

Libor Kvitek*, Robert Prucek, Ales Panacek and Jana Soukupova
RCPTM and Department of Physical Chemistry, Faculty of Science, Palacky University in Olomouc, Czech Republic

*Address all correspondence to: libor.kvitek@upol.cz

References

[1] Tadros TF. Industrial applications of dispersions. Advances in Colloid and Interface Science. 1993;**46**:1-47

[2] Rusk P, editor. Metal Nanoparticles: Concepts and Applications. New York: Willford Press; 2016

[3] Harish KK, Venkatesh N, Bhowmik H, Kuila A. Metallic nanoparticle: A review. Biomedical Journal of Scientific & Technical Research. 2018;**4**(2):3765-3775

[4] Creighton JA, Eadon DG. Ultraviolet visible absorption-spectra of the colloidal metallic elements. Journal of the Chemical Society, Faraday Transactions. 1991;**87**(24):3881-3891

[5] Scholl JA, Koh AL, Dionne JA. Quantum plasmon resonances of individual metallic nanoparticles. Nature. 2012;**483**(7390):421-468

[6] Ringe E, Sharma B, Henry AI, Marks LD, Van Duyne RP. Single nanoparticle plasmonics. Physical Chemistry Chemical Physics. 2013;**15**(12):4110-4129

[7] Marhaba S. Effect of size, shape and environment on the optical response of metallic nanoparticles. In: Seehra M, Bristow A, editors. Noble and Precious Metals. London: IntechOpen; 2017. pp. 215-238

[8] Araujo TP, Quiroz J, Barbosa ECM, Camargo PHC. Understanding plasmonic catalysis with controlled nanomaterials based on catalytic and plasmonic metals. Current Opinion in Colloid & Interface Science. 2019;**39**:110-122

[9] Ujihara M. Solution-phase synthesis of branched metallic nanoparticles for plasmonic applications. Journal of Oleo Science. 2018;**67**(6):689-696

[10] Eremina OE, Semenova AA, Sergeeva EA, Brazhe NA, Maksimov GV, Shekhovtsova TN, et al. Surface-enhanced Raman spectroscopy in modern chemical analysis: Advances and prospects. Russian Chemical Reviews. 2018;**87**(8):741-770

[11] Xing LX, Xiahou YJ, Zhang PN, Du W, Xia HB. Size control synthesis of monodisperse, quasi-spherical silver nanoparticles to realize surface-enhanced Raman scattering uniformity and reproducibility. ACS Applied Materials & Interfaces. 2019;**11**(19):17637-17646

[12] Sarfo DK, Izake EL, O'Mullane AP, Ayoko GA. Fabrication of nanostructured SERS substrates on conductive solid platforms for environmental application. Critical Reviews in Environmental Science and Technology. 2019;**49**(14):1294-1329

[13] Turzhitsky V, Qiu L, Itzkan I, Novikov AA, Kotelev MS, Getmanskiy M, et al. Spectroscopy of scattered light for the characterization of micro and nanoscale objects in biology and medicine. Applied Spectroscopy. 2014;**68**(2):133-154

[14] Sarycheva AS, Brazhe NA, Baizhumanov AA, Nikelshparg EI, Semenova AA, Garshev AV, et al. New nanocomposites for SERS studies of living cells and mitochondria. Journal of Materials Chemistry B. 2016;**4**(3):539-546

[15] Gong LS, Wang YP, Liu JB. Bioapplications of renal-clearable luminescent metal nanoparticles. Biomaterials Science. 2017;**5**(8): 1393-1406

[16] Klebowski B, Depciuch J, Parlinska-Wojtan M, Baran J. Applications of noble metal-based nanoparticles in medicine. International Journal of Molecular Sciences. 2018;**19**(12):4031

[17] Elahi N, Kamali M, Baghersad MH. Recent biomedical applications of gold nanoparticles: A review. Talanta. 2018;**184**:537-556

[18] Astruc D. Nanoparticles and Catalysis. Weinheim: Wiley-VCH; 2008

[19] Chaturvedi S, Dave PN, Shah NK. Applications of nano-catalyst in new era. Journal of Saudi Chemical Society. 2012;**16**(3):307-325

[20] Zhao PX, Feng XW, Huang DS, Yang GY, Astruc D. Basic concepts and recent advances in nitrophenol reduction by gold- and other transition metal nanoparticles. Coordination Chemistry Reviews. 2015;**287**:114-136

[21] Gubin SP. Magnetic nanoparticles. Wiley-VCH: Weinheim; 2009

[22] Kudr J, Haddad Y, Richtera L, Heger Z, Cernak M, Adam V, et al. Magnetic nanoparticles: From design and synthesis to real world applications. Nanomaterials. 2017;7(9):243

[23] Kvitek L, Panacek A, Soukupova J, Kolar M, Vecerova R, Prucek R, et al. Effect of surfactants and polymers on stability and antibacterial activity of silver nanoparticles (NPs). Journal of Physical Chemistry C. 2008;**112**(15):5825-5834

[24] Duran N, Duran M, de Jesus MB, Seabra AB, Favaro WJ, Nakazato G. Silver nanoparticles: A new view on mechanistic aspects on antimicrobial activity. Nanomedicine: Nanotechnology, Biology and Medicine. 2016;**12**(3):789-799

[25] Prasher P, Singh M, Mudila H. Silver nanoparticles as antimicrobial therapeutics: Current perspectives and future challenges. 3. Biotech. 2018;**14**:8(10)

[26] Panacek A, Kolar M, Vecerova R, Prucek R, Soukupova J, Krystof V, et al. Antifungal activity of silver nanoparticles against Candida spp. Biomaterials. 2009;**30**(31):6333-6340

[27] Kumar SSD, Rajendran NK, Houreld NN, Abrahamse H. Recent advances on silver nanoparticle and biopolymer-based biomaterials for wound healing applications. International Journal of Biological Macromolecules. 2018;**115**:165-175

[28] Deshmukh SP, Patil SM, Mullani SB, Delekar SD. Silver nanoparticles as an effective disinfectant: A review. Materials Science & Engineering, C: Materials for Biological Applications. 2019;**97**:954-965

[29] Auffan M, Rose J, Bottero JY, Lowry GV, Jolivet JP, Wiesner MR. Towards a definition of inorganic nanoparticles from an environmental, health and safety perspective. Nature Nanotechnology. 2009;**4**(10):634-641

[30] Panacek A, Kvitek L, Prucek R, Kolar M, Vecerova R, Pizurova N, et al. Silver colloid nanoparticles: Synthesis, characterization, and their antibacterial activity. The Journal of Physical Chemistry. B. 2006;**110**(33):16248-16253

[31] Agnihotri S, Mukherji S, Mukherji S. Size-controlled silver nanoparticles synthesized over the range 5-100 nm using the same protocol and their antibacterial efficacy. RSC Advances. 2014;**4**(8):3974-3983

[32] Ginjupalli K, Shaw T, Tellapragada C, Alla R, Gupta L, Perampalli NU. Does the size matter? Evaluation of effect of incorporation of silver nanoparticles of varying particle size on the antimicrobial activity and properties of irreversible hydrocolloid impression material. Dental Materials. 2018;**34**(7):E158-E165

[33] Njoki PN, Lim S, Mott D, Park HY, Khan B, Mishra S, et al. Size correlation of optical and spectroscopic

properties for gold nanoparticles. Journal of Physical Chemistry C. 2007;**111**(40):14664-14669

[34] Schneider S, Halbig P, Grau H, Nickel U. Reproducible preparation of silver sols with uniform particle-size for application in surface-enhanced Raman-spectroscopy. Photochemistry and Photobiology. 1994;**60**(6):605-610

[35] Emory SR, Haskins WE, Nie SM. Direct observation of size-dependent optical enhancement in single metal nanoparticles. Journal of the American Chemical Society. 1998;**120**(31):8009-8010

[36] Mogensen KB, Kneipp K. Size-dependent shifts of Plasmon resonance in silver nanoparticle films using controlled dissolution: Monitoring the onset of surface screening effects. Journal of Physical Chemistry C. 2014;**118**(48):28075-28083

[37] Sharma RK, Sharma P, Maitra A. Size-dependent catalytic behavior of platinum nanoparticles on the hexacyanoferrate(III)/thiosulfate redox reaction. Journal of Colloid and Interface Science. 2003;**265**(1):134-140

[38] Panigrahi S, Basu S, Praharaj S, Pande S, Jana S, Pal A, et al. Synthesis and size-selective catalysis by supported gold nanoparticles: Study on heterogeneous and homogeneous catalytic process. Journal of Physical Chemistry C. 2007;**111**(12):4596-4605

[39] Roldan CB. Synthesis and catalytic properties of metal nanoparticles: Size, shape, support, composition, and oxidation state effects. Thin Solid Films. 2010;**518**(12):3127-3150

[40] Panacek A, Prucek R, Hrbac J, Nevecna T, Steffkova J, Zboril R, et al. Polyacrylate-assisted size control of silver nanoparticles and their catalytic activity. Chemistry of Materials. 2014;**26**(3):1332-1339

[41] Roldan Cuenya B, Behafarid F. Nanocatalysis: Size- and shape-dependent chemisorption and catalytic reactivity. Surface Science Reports. 2015;**70**(2):135-187

[42] Suchomel P, Kvitek L, Prucek R, Panacek A, Halder A, Vajda S, et al. Simple size-controlled synthesis of Au nanoparticles and their size-dependent catalytic activity. Scientific Reports. 2018;**15**:8

[43] Carpenter EE, Seip CT, O'Connor CJ. Magnetism of nanophase metal and metal alloy particles formed in ordered phases. Journal of Applied Physics. 1999;**85**(8):5184-5186

[44] Rudakov GA, Tsiberkin KB, Ponomarev RS, Henner VK, Ziolkowska DA, Jasinski JB, et al. Magnetic properties of transition metal nanoparticles enclosed in carbon nanocages. Journal of Magnetism and Magnetic Materials. 2019;**472**:34-39

[45] Ivanova OS, Zamborini FP. Electrochemical size discrimination of gold nanoparticles attached to glass/indium-tin-oxide electrodes by oxidation in bromide-containing electrolyte. Analytical Chemistry. 2010;**82**(13):5844-5850

[46] Scanlon MD, Peljo P, Mendez MA, Smirnov E, Girault HH. Charging and discharging at the nanoscale: Fermi level equilibration of metallic nanoparticles. Chemical Science. 2015;**6**(5):2705-2720

[47] Letellier P, Mayaffre A, Turmine M. Melting point depression of nanosolids: Nonextensive thermodynamics approach. Physical Review B. 2007;**76**(4):045428

[48] Goswami GK, Nanda KK. Thermodynamic models for the size-dependent melting of nanoparticles: Different hypotheses. Current Nanoscience. 2012;**8**(2):305-311

[49] Gao F, Gu Z. Melting temperature of metallic nanoparticles. In: Aliofkhazraei M, editor. Handbook of Nanoparticles. Cham: Springer; 2016. pp. 661-690

[50] Wang BX, Zhou LP, Peng XF. Surface and size effects on the specific heat capacity of nanoparticles. International Journal of Thermophysics. 2006;**27**(1):139-151

[51] Likhachev VN, Vinogradov GA, Alymov MI. Anomalous heat capacity of nanoparticles. Physics Letters A. 2006;**357**(3):236-239

[52] Singh M, Lara S, Tlali S. Effects of size and shape on the specific heat, melting entropy and enthalpy of nanomaterials. Journal of Taibah University for Science. 2017;**11**(6):922-929

[53] Aiken JD, Finke RG. A review of modern transition-metal nanoclusters: Their synthesis, characterization, and applications in catalysis. Journal of Molecular Catalysis a-Chemical. 1999;**145**(1-2):1-44

[54] Mathew A, Pradeep T. Noble metal clusters: Applications in energy, environment, and biology. Particle and Particle Systems Characterization. 2014;**31**(10):1017-1053

[55] Campos A, Troc N, Cottancin E, Pellarin M, Weissker HC, Lerme J, et al. Plasmonic quantum size effects in silver nanoparticles are dominated by interfaces and local environments. Nature Physics. 2019;**15**(3):275

[56] Scholl JA, Garcia-Etxarri A, Koh AL, Dionne JA. Observation of quantum tunneling between two plasmonic nanoparticles. Nano Letters. 2013;**13**(2):564-569

[57] Scholl JA, Garcia-Etxarri A, Aguirregabiria G, Esteban R, Narayan TC, Koh AL, et al. Evolution of plasmonic metamolecule modes in the quantum tunneling regime. ACS Nano. 2016;**10**(1):1346-1354

[58] Zhang LB, Wang EK. Metal nanoclusters: New fluorescent probes for sensors and bioimaging. Nano Today. 2014;**9**(1):132-157

[59] Khandelwal P, Poddar P. Fluorescent metal quantum clusters: An updated overview of the synthesis, properties, and biological applications. Journal of Materials Chemistry B. 2017;**5**(46):9055-9084

[60] Jeseentharani V, Pugazhenthiran N, Mathew A, Chakraborty I, Baksi A, Ghosh J, et al. Atomically precise noble metal clusters harvest visible light to produce energy. ChemistrySelect. 2017;**2**(4):1454-1463

[61] Crawford SE, Hartmann MJ, Millstone JE. Surface chemistry-mediated near-infrared emission of small coinage metal nanoparticles. Accounts of Chemical Research. 2019;**52**(3):695-703

[62] Yu HZ, Rao B, Jiang W, Yang S, Zhu MZ. The photoluminescent metal nanoclusters with atomic precision. Coordination Chemistry Reviews. 2019;**378**:595-617

[63] Wang QY, Wang SY, Hu X, Li FY, Ling DS. Controlled synthesis and assembly of ultra-small nanoclusters for biomedical applications. Biomaterials Science. 2019;**7**(2):480-489

[64] Mittemeijer EJ. Fundamentals of Materials Science : The Microstructure-Property Relationship Using Metals as Model Systems. Heidelberg; New York: Springer; 2010

[65] Lin PC, Lin S, Wang PC, Sridhar R. Techniques for physicochemical characterization of nanomaterials. Biotechnology Advances. 2014;**32**(4):711-726

[66] Joy DC, Romig AD, Goldstein J. Principles of Analytical Electron Microscopy. New York: Plenum Press; 1986

[67] Fultz B, Howe JM. Transmission Electron Microscopy and Diffractometry of Materials. Heidelberg; New York: Springer; 2013

[68] Tanaka N. Scanning Transmission Electron Microscopy of Nanomaterials : Basics of Imaging and Analysis. London: Imperial College Press; 2014

[69] Giessibl FJ. Advances in atomic force microscopy. Reviews of Modern Physics. 2003;**75**(3):949-983

[70] Humphris ADL, Miles MJ, Hobbs JK. A mechanical microscope: High-speed atomic force microscopy. Applied Physics Letters. 2005;**86**(3):34106

[71] Gentry ST, Kendra SF, Bezpalko MW. Ostwald ripening in metallic nanoparticles: Stochastic kinetics. Journal of Physical Chemistry C. 2011;**115**(26):12736-12741

[72] Prucek R, Panacek A, Fargasova A, Ranc V, Masek V, Kvitek L, et al. Re-crystallization of silver nanoparticles in a highly concentrated NaCl environment-a new substrate for surface enhanced IR-visible Raman spectroscopy. CrystEngComm. 2011;**13**(7):2242-2248

[73] Panacek A, Kvitek L, Smekalova M, Vecerova R, Kolar M, Roderova M, et al. Bacterial resistance to silver nanoparticles and how to overcome it. Nature Nanotechnology. 2018;**13**(1):65-71

[74] Roh J, Yi J, Kim Y. Rapid, reversible preparation of size-controllable silver nanoplates by chemical redox. Langmuir. 2010;**26**(14):11621-11623

[75] Berne BJ, Pecora R. Dynamic Light Scattering : With Applications to Chemistry, Biology, and Physics. Dover ed. Mineola, NY: Dover Publications; 2000

[76] Bhattacharjee S. DLS and zeta potential–What they are and what they are not? Journal of Controlled Release. 2016;**235**:337-351

[77] Maguire CM, Rosslein M, Wick P, Prina-Mello A. Characterisation of particles in solution–A perspective on light scattering and comparative technologies. Science and Technology of Advanced Materials. 2018;**19**(1):732-745

[78] Allen T. Powder Sampling and Particle Size Determination. 1st ed. Amsterdam; Boston: Elsevier; 2003

[79] Hagendorfer H, Kaegi R, Parlinska M, Sinnet B, Ludwig C, Ulrich A. Characterization of silver nanoparticle products using asymmetric flow field flow fractionation with a multidetector approach–A comparison to transmission electron microscopy and batch dynamic light scattering. Analytical Chemistry. 2012;**84**(6):2678-2685

[80] Bootz A, Vogel V, Schubert D, Kreuter J. Comparison of scanning electron microscopy, dynamic light scattering and analytical ultracentrifugation for the sizing of poly(butyl cyanoacrylate) nanoparticles. European Journal of Pharmaceutics and Biopharmaceutics. 2004;**57**(2):369-375

[81] Mahl D, Diendorf J, Meyer-Zaika W, Epple M. Possibilities and limitations of different analytical methods for the size determination of a bimodal dispersion of metallic nanoparticles. Colloids and Surfaces A-Physicochemical and Engineering Aspects. 2011;**377**(1-3):386-392

[82] Boyd RD, Pichaimuthu SK, Cuenat A. New approach to inter-technique comparisons for nanoparticle size measurements; using atomic

force microscopy, nanoparticle tracking analysis and dynamic light scattering. Colloids and Surfaces A-Physicochemical and Engineering Aspects. 2011;**387**(1-3):35-42

[83] Souza TGF, Ciminelli VST, Mohallem NDS. A comparison of TEM and DLS methods to characterize size distribution of ceramic nanoparticles. In: 8th Brazilian Congress on Metrology (Metrologia 2015). 2016. p. 733

[84] Hallett FR, Watton J, Krygsman P. Vesicle sizing–Number distributions by dynamic light-scattering. Biophysical Journal. 1991;**59**(2):357-362

[85] Vorokh AS. Rrer formula: Estimation of error in determining small nanoparticle size. Nanosystems: Physics, Chemistry, Mathematics. 2018;**9**(3):364-369

[86] Bubb DM, O'Malley SM, Schoeffling J, Jimenez R, Zinderman B, Yi SY. Size control of gold nanoparticles produced by laser ablation of thin films in an aqueous environment. Chemical Physics Letters. 2013;**565**:65-68

[87] Zhang JM, Claverie J, Chaker M, Ma DL. Colloidal metal nanoparticles prepared by laser ablation and their applications. ChemPhysChem. 2017;**18**(9):986-1006

[88] Zhang JM, Chaker M, Ma DL. Pulsed laser ablation based synthesis of colloidal metal nanoparticles for catalytic applications. Journal of Colloid and Interface Science. 2017;**489**:138-149

[89] Sportelli MC, Izzi M, Volpe A, Clemente M, Picca RA, Ancona A, et al. The pros and cons of the use of laser ablation synthesis for the production of silver nano-antimicrobials. Antibiotics-Basel. 2018;**7**(3):E67

[90] Yan ZJ, Chrisey DB. Pulsed laser ablation in liquid for micro−/ nanostructure generation. Journal of Photochemistry and Photobiology, C: Photochemistry Reviews. 2012;**13**(3):204-223

[91] Baruah PK, Sharma AK, Khare A. Effective control of particle size, surface plasmon resonance and stoichiometry of Cu@CuxO nanoparticles synthesized by laser ablation of Cu in distilled water. Optics and Laser Technology. 2018;**108**:574-582

[92] Baruah PK, Singh A, Rangan L, Sharma AK, Khare A. Optimization of copper nanoparticles synthesized by pulsed laser ablation in distilled water as a viable SERS substrate for karanjin. Materials Chemistry and Physics. 2018;**220**:111-117

[93] Wang ZY, Zhou R, Wen F, Zhang RK, Ren L, Teoh SH, et al. Reliable laser fabrication: The quest for responsive biomaterials surface. Journal of Materials Chemistry B. 2018;**6**(22):3612-3631

[94] Zeng HB, Du XW, Singh SC, Kulinich SA, Yang SK, He JP, et al. Nanomaterials via laser ablation/ irradiation in liquid: A review. Advanced Functional Materials. 2012;**22**(7):1333-1353

[95] Gonzalez-Rubio G, Guerrero-Martinez A, Liz-Marzan LM. Reshaping, fragmentation, and assembly of gold nanoparticles assisted by pulse lasers. Accounts of Chemical Research. 2016;**49**(4): 678-686

[96] Mafune F, Kohno J, Takeda Y, Kondow T, Sawabe H. Formation and size control of silver nanoparticles by laser ablation in aqueous solution. The Journal of Physical Chemistry. B. 2000;**104**(39):9111-9117

[97] Mafune F, Kohno J, Takeda Y, Kondow T, Sawabe H. Formation of gold nanoparticles by laser ablation in aqueous solution of surfactant.

The Journal of Physical Chemistry. B. 2001;**105**(22):5114-5120

[98] Fan GH, Ren ST, Qu SL, Wang Q, Gao RX, Han M. Stability and nonlinear optical properties of Cu nanoparticles prepared by femtosecond laser ablation of Cu target in alcohol and water. Optics Communication. 2014;**330**:122-130

[99] Magnusson MH, Deppert K, Malm JO, Bovin JO, Samuelson L. Gold nanoparticles: Production, reshaping, and thermal charging. Journal of Nanoparticle Research. 1999;**1**(2): 243-251

[100] Kruis FE, Fissan H, Rellinghaus B. Sintering and evaporation characteristics of gas-phase synthesis of size-selected PbS nanoparticles. Materials Science & Engineering, B: Solid-State Materials for Advanced Technology. 2000;**69**: 329-334

[101] Rahman MM, Hattori N, Nakagawa Y, Lin X, Yagai S, Sakai M, et al. Preparation and characterization of silver nanoparticles on localized surface plasmon-enhanced optical absorption. Japanese Journal of Applied Physics. 2014;**53**(11):11S

[102] Kim HR, Sahu BB, Xiang PJ, Han JG. Direct synthesis of magnetron sputtered nanostructured Cu films with desired properties via plasma chemistry for their efficient antibacterial application. Plasma Processes and Polymers. 2018;**15**(9):1800009

[103] Hirsch UM, Teuscher N, Ruhl M, Heilmann A. Plasma-enhanced magnetron sputtering of silver nanoparticles on reverse osmosis membranes for improved antifouling properties. Surfaces and Interfaces. 2019;**16**:1-7

[104] Yin GL, Bai SH, Tu XL, Li Z, Zhang YP, Wang WM, et al. Highly sensitive and stable SERS substrate fabricated by co-sputtering and atomic layer deposition. Nanoscale Research Letters. 2019;**18**:14

[105] Siegel J, Kvitek O, Ulbrich P, Kolska Z, Slepicka P, Svorcik V. Progressive approach for metal nanoparticle synthesis. Materials Letters. 2012;**89**:47-50

[106] Wender H, Migowski P, Feil AF, Teixeira SR, Dupont J. Sputtering deposition of nanoparticles onto liquid substrates: Recent advances and future trends. Coordination Chemistry Reviews. 2013;**257**(17-18):2468-2483

[107] Nguyen MT, Yonezawa T. Sputtering onto a liquid: Interesting physical preparation method for multi-metallic nanoparticles. Science and Technology of Advanced Materials. 2018;**19**(1):883-898

[108] Varadan VK. Nanoscience and Nanotechnology in Engineering. Singapore; Hackensack, NJ: World Scientific; 2010

[109] Lamer VK, Dinegar RH. Theory, production and mechanism of formation of monodispersed hydrosols. Journal of the American Chemical Society. 1950;**72**(11):4847-4854

[110] Mandal M, Ghosh SK, Kundu S, Esumi K, Pal T. UV photoactivation for size and shape controlled synthesis and coalescence of gold nanoparticles in micelles. Langmuir. 2002;**18**(21):7792-7797

[111] Murshid N, Smith DS, Kitaev V. Photochemical formation of tunable gold nanostructures using versatile water-soluble thiolate Au(I) precursor. Particle & Particle Systems Characterization. 2018;**35**(11):1800285

[112] Bartlett TR, Sokolov SV, Compton RG. Nanoparticle photochemistry via nano-impacts. Russian Journal of Electrochemistry. 2016;**52**(12):1131-1136

[113] Stamplecoskie KG, Scaiano JC. Silver as an example of the applications of photochemistry to the synthesis and uses of nanomaterials. Photochemistry and Photobiology. 2012;**88**(4):762-768

[114] Abedini A, Daud AR, Hamid MAA, Othman NK, Saion E. A review on radiation-induced nucleation and growth of colloidal metallic nanoparticles. Nanoscale Research Letters. 2013;**13**:8

[115] Abedini A, Bakar AAA, Larki F, Menon PS, Islam MS, Shaari S. Recent advances in shape-controlled synthesis of noble metal nanoparticles by radiolysis route. Nanoscale Research Letters. 2016;**11**:1-13

[116] El-Batal AI, Mosallam FM, El-Sayyad GS. Synthesis of metallic silver nanoparticles by fluconazole drug and gamma rays to inhibit the growth of multidrug-resistant microbes. Journal of Cluster Science. 2018;**29**(6):1003-1015

[117] Wang M, Park C, Woehl TJ. Quantifying the nucleation and growth kinetics of electron beam nanochemistry with liquid cell scanning transmission electron microscopy. Chemistry of Materials. 2018;**30**(21):7727-7736

[118] ThiteAG,KrishnanandK,SharmaDK, Mukhopadhyay AK. Multifunctional finishing of cotton fabric by electron beam radiation synthesized silver nanoparticles. Radiation Physics and Chemistry. 2018;**153**:173-179

[119] Vais RD, Sattarahmady N, Heli H. Green electrodeposition of gold nanostructures by diverse size, shape, and electrochemical activity. Gold Bulletin. 2016;**49**(3-4):95-102

[120] Yang SM, Paranthaman MP, Noh TW, Kalinin SV, Strelcov E. Nanoparticle shape evolution and proximity effects during tip-induced electrochemical processes. ACS Nano. 2016;**10**(1):663-671

[121] Brasiliense V, Clausmeyer J, Dauphin AL, Noel JM, Berto P, Tessier G, et al. Opto-electrochemical in situ monitoring of the cathodic formation of single cobalt nanoparticles. Angewandte Chemie, International Edition. 2017;**56**(35):10598-10601

[122] Mandke MV, Pathan HM. Electrochemical growth of copper nanoparticles: Structural and optical properties. Journal of Electroanalytical Chemistry. 2012;**686**:19-24

[123] Nishimura T, Nakade T, Morikawa T, Inoue H. Effect of current density on electrochemical shape control of Pt nanoparticles. Electrochimica Acta. 2014;**129**:152-159

[124] Zayer MQ, Alwan AM, Ahmed AS, Dheyab AB. Accurate controlled deposition of silver nanoparticles on porous silicon by drifted ions in electrolytic solution. Current Applied Physics. 2019;**19**(9):1024-1030

[125] Bian JC, Li Z, Chen ZD, He HY, Zhang XW, Li X, et al. Electrodeposition of silver nanoparticle arrays on ITO coated glass and their application as reproducible surface-enhanced Raman scattering substrate. Applied Surface Science. 2011;**258**(5):1831-1835

[126] Grez P, Rojas C, Segura I, Heyser C, Ballesteros L, Celedon C, et al. Photoelectrochemical properties of nanostructured copper oxides formed sonoelectrochemically. International Journal of Electrochemical Science. 2017;**12**(8):7240-7248

[127] Yanilkin VV, Nastapova NV, NasretdinovaGR,FedorenkoSV,JilkinME, Mustafina AR, et al. Methylviologen mediated electrosynthesis of gold nanoparticles in the solution bulk. RSC Advances. 2016;**6**(3):1851-1859

[128] Nasretdinova GR, Osin YN, Gubaidullin AT, Yanilkin VV. Methylviologen mediated electrosynthesis of palladium nanoparticles stabilized with CTAC. Journal of the Electrochemical Society. 2016;**163**(8):G99-G106

[129] Kokorekin VA, Gamayunova AV, Yanilkin VV, Petrosyan VA. Mediated electrochemical synthesis of copper nanoparticles in solution. Russian Chemical Bulletin. 2017;**66**(11):2035-2043

[130] Yanilkin VV, Nastapova NV, Nasretdinova GR, Osin YN, Evtjugin VG, Ziganshina AY, et al. Structure and catalytic activity of ultrasmall Rh, Pd and (Rh plus Pd) nanoparticles obtained by mediated electrosynthesis. New Journal of Chemistry. 2019;**43**(9):3931-3945

[131] Abou El-Nour KMM, Eftaiha A, Al-Warthan A, Ammar RAA. Synthesis and applications of silver nanoparticles. Arabian Journal of Chemistry. 2010;**3**(3):135-140

[132] Guo SJ, Wang EK. Noble metal nanomaterials: Controllable synthesis and application in fuel cells and analytical sensors. Nano Today. 2011;**6**(3):240-264

[133] Khan M, Shaik MR, Adil SF, Khan ST, Al-Warthan A, Siddiqui MRH, et al. Plant extracts as green reductants for the synthesis of silver nanoparticles: Lessons from chemical synthesis. Dalton Transactions. 2018;**47**(35):11988-12010

[134] Zhang Z, Shen WF, Xue J, Liu YM, Liu YW, Yan PP, et al. Recent advances in synthetic methods and applications of silver nanostructures. Nanoscale Research Letters. 2018;**18**:13

[135] Fievet F, Ammar-Merah S, Brayner R, Chau F, Giraud M, Mammeri F, et al. The polyol process: A unique method for easy access to metal nanoparticles with tailored sizes, shapes and compositions. Chemical Society Reviews. 2018;**47**(14):5187-5233

[136] Atkins PW, De Paula J. Elements of Physical Chemistry. 6th ed. Oxford: Oxford University Press; 2013

[137] Masitas RA, Zamborini FP. Oxidation of highly unstable < 4 nm diameter gold nanoparticles 850 mV negative of the bulk oxidation potential. Journal of the American Chemical Society. 2012;**134**(11):5014-5017

[138] Troupis A, Triantis T, Hiskia A, Papaconstantinou E. Rate-redox-controlled size-selective synthesis of silver nanoparticles using polyoxometalates. European Journal of Inorganic Chemistry. 2008;**36**:5579-5586

[139] Haynes WM, Lide DR, Bruno TJ. CRC Handbook of Chemistry and Physics : A Ready-Reference Book of Chemical and Physical Data. 96th ed. Boca Raton, FL; London: CRC Press; 2015

[140] Do D. Electrochemical Data : A Handbook for Electrochemists in Industry and Universities. Amsterdam; New York: Elsevier Scientific Pub. Co.; 1975

[141] Quester K, Avalos-Borja M, Castro-Longoria E. Biosynthesis and microscopic study of metallic nanoparticles. Micron. 2013;**54-55**:1-27

[142] Adil SF, Assal ME, Khan M, Al-Warthan A, Siddiqui MRH, Liz-Marzan LM. Biogenic synthesis of metallic nanoparticles and prospects toward green chemistry. Dalton Transactions. 2015;**44**(21):9709-9717

[143] Beyene HD, Werkneh AA, Bezabh HK, Ambaye TG. Synthesis paradigm and applications of silver nanoparticles (AgNPs), a review. Sustainable Materials and Technologies. 2017;**13**:18-23

[144] Khandel P, Yadaw RK, Soni DK, Kanwar L, Shahi SK. Biogenesis of metal nanoparticles and their pharmacological applications: Present status and application prospects. Journal of Nanostructure in Chemistry. 2018;**8**(3):217-254

[145] Ali J, Ali N, Wang L, Waseem H, Pan G. Revisiting the mechanistic pathways for bacterial mediated synthesis of noble metal nanoparticles. Journal of Microbiological Methods. 2019;**159**:18-25

[146] Gunarani GI, Raman AB, Kumar JD, Natarajan S, Jegadeesan GB. Biogenic synthesis of Fe and NiFe nanoparticles using *Terminalia bellirica* extracts for water treatment applications. Materials Letters. 2019;**247**:90-94

[147] Ravikumar KVG, Sudakaran SV, Ravichandran K, Pulimi M, Natarajan C, Mukherjee A. Green synthesis of NiFe nano particles using *Punica granatum* peel extract for tetracycline removal. Journal of Cleaner Production. 2019;**210**:767-776

[148] Costentin C, Louault C, Robert M, Saveant JM. The electrochemical approach to concerted proton-electron transfers in the oxidation of phenols in water. Proceedings of the National Academy of Sciences of the United States of America. 2009;**106**(43):18143-18148

[149] Polak J. Structure/redox potential relationship of simple organic compounds as potential precursors of dyes for laccase-mediated transformation. Biotechnology Progress. 2012;**28**(1):93-102

[150] Rodrigues TS, Zhao M, Yang TH, Gilroy KD, da Silva AGM, Camargo PHC, et al. Synthesis of colloidal metal nanocrystals: A comprehensive review on the reductants. Chemistry—A European Journal. 2018;**24**(64):16944-16963

[151] Holade Y, Engel AB, Servat K, Napporn TW, Morais C, Tingry S, et al. Electrocatalytic and electroanalytic investigation of carbohydrates oxidation on gold-based nanocatalysts in alkaline and neutral pHs. Journal of the Electrochemical Society. 2018;**165**(9):H425-H436

[152] Roto R, Rasydta HP, Suratman A, Aprilita NH. Effect of reducing agents on physical and chemical properties of silver nanoparticles. Indonesian Journal of Chemistry. 2018;**18**(4):614-620

[153] Creighton JA, Blatchford CG, Albrecht MG. Plasma resonance enhancement of Raman-scattering by pyridine adsorbed on silver or gold sol particles of size comparable to the excitation wavelength. Journal of the Chemical Society-Faraday Transactions. 1979;**75**:790-798

[154] Philip D. Synthesis and spectroscopic characterization of gold nanoparticles. Spectrochimica Acta Part A-Molecular and Biomolecular Spectroscopy. 2008;**71**(1):80-85

[155] Jiang XC, Chen CY, Chen WM, Yu AB. Role of citric acid in the formation of silver nanoplates through a synergistic reduction approach. Langmuir. 2010;**26**(6):4400-4408

[156] Pinto VV, Ferreira MJ, Silva R, Santos HA, Silva F, Pereira CM. Long time effect on the stability of silver nanoparticles in aqueous medium: Effect of the synthesis and storage conditions. Colloids and Surfaces A-Physicochemical and Engineering Aspects. 2010;**364**(1-3):19-25

[157] Lee PC, Meisel D. Adsorption and surface-enhanced Raman of dyes on silver and gold sols. The Journal of Physical Chemistry. 1982;**86**(17):3391-3395

[158] Larmour IA, Faulds K, Graham D. SERS activity and stability of the

most frequently used silver colloids. Journal of Raman Spectroscopy. 2012;**43**(2):202-206

[159] Mikac L, Ivanda M, Gotic M, Mihelj T, Horvat L. Synthesis and characterization of silver colloidal nanoparticles with different coatings for SERS application. Journal of Nanoparticle Research. 2014;**16**(12):2748-1

[160] Turkevich J, Stevenson PC, Hillier J. A study of the nucleation and growth processes in the synthesis of colloidal gold. Discussions of the Faraday Society. 1951;**11**:55

[161] Turkevich J, Stevenson PC, Hillier J. The formation of colloidal gold. The Journal of Physical Chemistry. 1953;**57**(7):670-673

[162] Enustun BV, Turkevich J. Coagulation of colloidal gold. Journal of the American Chemical Society. 1963;**85**(21):3317-3328

[163] Kvitek L, Prucek R, Panacek A, Novotny R, Hrbac J, Zboril R. The influence of complexing agent concentration on particle size in the process of SERS active silver colloid synthesis. Journal of Materials Chemistry. 2005;**15**(10):1099-1105

[164] Michalcova A, Machado L, Marek I, Martinec M, Slukova M, Vojtech D. Properties of Ag nanoparticles prepared by modified Tollens' process with the use of different saccharide types. Journal of Physics and Chemistry of Solids. 2018;**113**:125-133

[165] Afshinnia K, Marrone B, Baalousha M. Potential impact of natural organic ligands on the colloidal stability of silver nanoparticles. The Science of the Total Environment. 2018;**625**:1518-1526

[166] Bhattarai B, Zaker Y, Bigioni TP. Green synthesis of gold and silver nanoparticles: Challenges and opportunities. Current Opinion in Green and Sustainable Chemistry. 2018;**12**:91-100

[167] Amin ZR, Khashyarmanesh Z, Bazzaz BSF, Noghabi ZS. Does biosynthetic silver nanoparticles are more stable with lower toxicity than their synthetic counterparts? Iranian Journal of Pharmaceutical Research. 2019;**18**(1):210-221

[168] Han J, Li LY, Guo R. Novel approach to controllable synthesis of gold nanoparticles supported on polyaniline nanofibers. Macromolecules. 2010;**43**(24):10636-10644

[169] Signori AM, Santos KD, Eising R, Albuquerque BL, Giacomelli FC, Domingos JB. Formation of catalytic silver nanoparticles supported on branched polyethyleneimine derivatives. Langmuir. 2010;**26**(22):17772-17779

[170] Han J, Wang MG, Hu YM, Zhou CQ, Guo R. Conducting polymer-noble metal nanoparticle hybrids: Synthesis mechanism application. Progress in Polymer Science. 2017;**70**:52-91

[171] Gregor L, Reilly AK, Dickstein TA, Mazhar S, Bram S, Morgan DG, et al. Facile synthesis of magnetically recoverable Pd and Ru catalysts for 4-nitrophenol reduction: Identifying key factors. Acs Omega. 2018;**3**(11):14717-14725

[172] Bouche M, Fournel S, Kichler A, Selvam T, Gallani JL, Bellemin-Laponnaz S. Straightforward synthesis of L-PEI-coated gold nanoparticles and their biological evaluation. European Journal of Inorganic Chemistry. 2018;**6**(25):2972-2975

[173] Batista CCS, Albuquerque LJC, de Araujo I, Albuquerque BL, da Silva FD, Giacomelli FC. Antimicrobial activity of nano-sized silver colloids stabilized by nitrogen-containing polymers: The key

influence of the polymer capping. RSC Advances. 2018;**8**(20):10873-10882

[174] Sarkar R, Kumar CA, Kumbhakar P, Mandal T. Aqueous synthesis and antibacterial activity of silver nanoparticles against pseudomonas putida. Materials Today-Proceedings. 2019;**11**:686-694

[175] Kotrlý S, Šůcha L. Handbook of Chemical Equilibria in Analytical Chemistry. Chichester; New York: Horwood; Halsted Press; 1985

[176] Yin YD, Li ZY, Zhong ZY, Gates B, Xia YN, Venkateswaran S. Synthesis and characterization of stable aqueous dispersions of silver nanoparticles through the Tollens process. Journal of Materials Chemistry. 2002;**12**(3):522-527

[177] Hussain JI, Talib A, Kumar S, Al-Thabaiti SA, Hashmi AA, Khan Z. Time dependence of nucleation and growth of silver nanoparticles. Colloids and Surfaces A-Physicochemical and Engineering Aspects. 2011;**381**(1-3):23-30

[178] Dondi R, Su W, Griffith GA, Clark G, Burley GA. Highly size- and shape-controlled synthesis of silver nanoparticles via a templated tollens reaction. Small. 2012;**8**(5):770-776

[179] Panneerselvam R, Xiao L, Waites KB, Atkinson TP, Dluhy RA. A rapid and simple chemical method for the preparation of Ag colloids for surface-enhanced Raman spectroscopy using the Ag mirror reaction. Vibrational Spectroscopy. 2018;**98**:1-7

[180] Prucek R, Panacek A, Soukupova J, Novotny R, Kvitek L. Reproducible synthesis of silver colloidal particles tailored for application in near-infrared surface-enhanced Raman spectroscopy. Journal of Materials Chemistry. 2011;**21**(17):6416-6420

[181] Evanoff DD, Chumanov G. Size-controlled synthesis of nanoparticles. 1. "silver-only" aqueous suspensions via hydrogen reduction. The Journal of Physical Chemistry. B. 2004;**108**(37):13948-13956

[182] Wiberg E, Wiberg N, Holleman AF. Inorganic Chemistry. 1st ed. San Diego; Berlin; New York: Academic Press; De Gruyter; 2001

[183] Leopold N, Lendl B. A new method for fast preparation of highly surface-enhanced Raman scattering (SERS) active silver colloids at room temperature by reduction of silver nitrate with hydroxylamine hydrochloride. The Journal of Physical Chemistry. B. 2003;**107**(24):5723-5727

[184] Nickel U, Mansyreff K, Schneider S. Production of monodisperse silver colloids by reduction with hydrazine: The effect of chloride and aggregation on SER(R) S signal intensity. Journal of Raman Spectroscopy. 2004;**35**(2):101-110

[185] Pal A, Shah S, Devi S. Synthesis of Au, Ag and Au-Ag alloy nanoparticles in aqueous polymer solution. Colloids and Surfaces A-Physicochemical and Engineering Aspects. 2007;**302**(1-3):51-57

[186] Puranik SS, Joshi HM, Ogale SB, Paknikar KM. Hydrazine based facile synthesis and ordered assembly of metal nanoparticles (Au, Ag) on a bacterial surface layer protein template. Journal of Nanoscience and Nanotechnology. 2008;**8**(7):3565-3569

[187] Minati L, Benetti F, Chiappini A, Speranza G. One-step synthesis of star-shaped gold nanoparticles. Colloids and Surfaces A-Physicochemical and Engineering Aspects. 2014;**441**:623-628

[188] Todor IS, Szabo L, Marisca OT, Chis V, Leopold N. Gold nanoparticle assemblies of controllable size obtained by hydroxylamine reduction at room

temperature. Journal of Nanoparticle Research. 2014;**16**(12):2740

[189] Cheng ZP, Chu XZ, Wu XQ, Xu JM, Zhong H, Yin JZ. Controlled synthesis of silver nanoplates and nanoparticles by reducing silver nitrate with hydroxylamine hydrochloride. Rare Metals. 2017;**36**(10):799-805

[190] Sucha L, Kotrly S. Solution Equilibria in Analytical Chemistry. London, New York: Van Nostrand Reinhold Co.; 1972

[191] Uchimiya M, Stone AT. Reversible redox chemistry of quinones: Impact on biogeochemical cycles. Chemosphere. 2009;**77**(4):451-458

[192] Huynh MT, Anson CW, Cavell AC, Stahl SS, Hammes-Schiffer S. Quinone 1 e(−) and 2 e(−)/2 H^+ reduction potentials: Identification and analysis of deviations from systematic scaling relationships. Journal of the American Chemical Society. 2016;**138**(49):15903-15910

[193] Martinez-Cifuentes M, Salazar R, Ramirez-Rodriguez O, Weiss-Lopez B, Araya-Maturana R. Experimental and theoretical reduction potentials of some biologically active ortho-carbonyl Para-quinones. Molecules. 2017;**22**(4):E557

[194] Jacob JA, Mahal HS, Biswas N, Mukherjee T, Kapoor S. Role of phenol derivatives in the formation of silver nanoparticles. Langmuir. 2008;**24**(2):528-533

[195] Xie T, Jing C, Li M, Ma W, Ding ZF, Long YT. pH-response mechanism of a redox reaction between silver ions and hydroquinone. Journal of Physical Chemistry C. 2016;**120**(40):23104-23110

[196] Qin YQ, Ji XH, Jing J, Liu H, Wu HL, Yang WS. Size control over spherical silver nanoparticles by ascorbic acid reduction. Colloids and Surfaces A-Physicochemical and Engineering Aspects. 2010;**372**(1-3):172-176

[197] Chen B, Jiao XL, Chen DR. Size-controlled and size-designed synthesis of nano/submicrometer Ag particles. Crystal Growth & Design. 2010;**10**(8):3378-3386

[198] Kozlova ES, Nikiforova TE. Incorporation of silver nanoparticles into a cellulose matrix for preparing package materials for foodstuffs. Russian Journal of Applied Chemistry. 2015;**88**(4):638-646

[199] Tang JQ, Fu XW, Ou QH, Gao KP, Man SQ, Guo J, et al. Hydroxide assisted synthesis of monodisperse and biocompatible gold nanoparticles with dextran. Materials Science & Engineering, C: Materials for Biological Applications. 2018;**93**:759-767

[200] Aguilar NM, Arteaga-Cardona F, Estevez JO, Silva-Gonzalez NR, Benitez-Serrano JC, Salazar-Kuri U. Controlled biosynthesis of silver nanoparticles using sugar industry waste, and its antimicrobial activity. Journal of Environmental Chemical Engineering. 2018;**6**(5):6275-6281

[201] Si S, Bhattacharjee RR, Banerjee A, Mandal TK. A mechanistic and kinetic study of the formation of metal nanoparticles by using synthetic tyrosine-based oligopeptides. Chemistry—A European Journal. 2006;**12**(4):1256-1265

[202] Palui G, Ray S, Banerjee A. Synthesis of multiple shaped gold nanoparticles using wet chemical method by different dendritic peptides at room temperature. Journal of Materials Chemistry. 2009;**19**(21):3457-3468

[203] Lah NAC, Johan MR. Facile shape control synthesis and optical properties of silver nanoparticles stabilized by Daxad 19 surfactant. Applied Surface Science. 2011;**257**(17):7494-7500

[204] Zhang AQ, Cai LJ, Sui L, Qian DJ, Chen M. Reducing properties of polymers in the synthesis of noble metal nanoparticles. Polymer Reviews. 2013;**53**(2):240-276

[205] Li YF, Gan WP, Zhou J, Lu ZQ, Yang C, Ge TT. Hydrothermal synthesis of silver nanoparticles in Arabic gum aqueous solutions. Transactions of Nonferrous Metals Society of China. 2015;**25**(6):2081-2086

[206] Biao LH, Tan SN, Wang YL, Guo XM, Fu YJ, Xu FJ, et al. Synthesis, characterization and antibacterial study on the chitosan-functionalized Ag nanoparticles. Materials Science & Engineering, C: Materials for Biological Applications. 2017;**76**:73-80

[207] Kora AJ, Sashidhar RB. Biogenic silver nanoparticles synthesized with rhamnogalacturonan gum: Antibacterial activity, cytotoxicity and its mode of action. Arabian Journal of Chemistry. 2018;**11**(3):313-323

[208] Zhang XW, Sun HY, Tan SN, Gao J, Fu YJ, Liu ZG. Hydrothermal synthesis of Ag nanoparticles on the nanocellulose and their antibacterial study. Inorganic Chemistry Communications. 2019;**100**:44-50

[209] Shou QH, Guo C, Yang LR, Jia LW, Liu CZ, Liu HZ. Effect of pH on the single-step synthesis of gold nanoparticles using PEO-PPO-PEO triblock copolymers in aqueous media. Journal of Colloid and Interface Science. 2011;**363**(2):481-489

[210] Merga G, Wilson R, Lynn G, Milosavljevic BH, Meisel D. Redox catalysis on "naked" silver nanoparticles. Journal of Physical Chemistry C. 2007;**111**(33):12220-12226

[211] Prucek R, Ranc V, Balzerova O, Panacek A, Zboril R, Kvitek L. Preparation of silver particles and its application for surface enhanced Raman scattering with near-infrared excitation. Materials Research Bulletin. 2014;**50**:63-67

[212] Agunloye E, Panariello L, Gavriilidis A, Mazzei L. A model for the formation of gold nanoparticles in the citrate synthesis method. Chemical Engineering Science. 2018;**191**:318-331

[213] Kettemann F, Birnbaum A, Witte S, Wuithschick M, Pinna N, Kraehnert R, et al. Missing piece of the mechanism of the Turkevich method: The critical role of citrate protonation. Chemistry of Materials. 2016;**28**(11):4072-4081

[214] Peckham GD, McNaught IJ. The variation of electrochemical cell potentials with temperature. Journal of Chemical Education. 2011;**88**(6):782-783

[215] Vekilov PG. The two-step mechanism of nucleation of crystals in solution. Nanoscale. 2010;**2**(11):2346-2357

[216] You HJ, Fang JX. Particle-mediated nucleation and growth of solution-synthesized metal nanocrystals: A new story beyond the LaMer curve. Nano Today. 2016;**11**(2):145-167

[217] Zhou Y, Wang HX, Lin WS, Lin LQ, Gao YX, Yang F, et al. Quantitative nucleation and growth kinetics of gold nanoparticles via model-assisted dynamic spectroscopic approach. Journal of Colloid and Interface Science. 2013;**407**:8-16

[218] Si S, Dinda E, Mandal TK. In situ synthesis of shape-selective gold nanocrystals using oligopeptide template: Effect of various reaction parameters. Journal of Nanoscience and Nanotechnology. 2008;**8**(11):5934-5941

[219] Nishimoto M, Abe S, Yonezawa T. Preparation of Ag nanoparticles using hydrogen peroxide as a reducing agent. New Journal of Chemistry. 2018;**42**(17):14493-14501

[220] Kvitek L, Bartova A, Prucek R, Panacek A, Soukupova J, Zboril R, editors. Low Temperature Synthesis of Silver Nanoparticles 10th Anniversary International Conference on Nanomaterials–Research and Application, NANOCON 2018; 2018; Brno, Czech Republic. Ostrava, Czech Republic: Tanger Ltd.; 2018

[221] Agunloye E, Gavriilidis A, Mazzei L. A mathematical investigation of the Turkevich organizer theory in the citrate method for the synthesis of gold nanoparticles. Chemical Engineering Science. 2017;**173**:275-286

[222] Shi L, Buhler E, Boue F, Carn F. How does the size of gold nanoparticles depend on citrate to gold ratio in Turkevich synthesis? Final answer to a debated question. Journal of Colloid and Interface Science. 2017;**492**:191-198

[223] Sardar R, Funston AM, Mulvaney P, Murray RW. Gold nanoparticles: Past, present, and future. Langmuir. 2009;**25**(24):13840-13851

[224] Mourdikoudis S, Liz-Marzan LM. Oleylamine in nanoparticle synthesis. Chemistry of Materials. 2013;**25**(9):1465-1476

[225] Ling DS. Surface ligands in synthesis, modification and assembly of nanoparticles for biomedical applications. Nanomedicine: Nanotechnology, Biology and Medicine. 2016;**12**(2):461-461

[226] Larm NE, Essner JB, Pokpas K, Canon JA, Jahed N, Iwuoha EI, et al. Room-temperature Turkevich method: Formation of gold nanoparticles at the speed of mixing using cyclic oxocarbon reducing agents. Journal of Physical Chemistry C. 2018;**122**(9):5105-5118

[227] Soukupova J, Kvitek L, Panacek A, Nevecna T, Zboril R. Comprehensive study on surfactant role on silver nanoparticles (NPs) prepared via modified Tollens process. Materials Chemistry and Physics. 2008;**111**(1):77-81

[228] Heinz H, Pramanik C, Heinz O, Ding YF, Mishra RK, Marchon D, et al. Nanoparticle decoration with surfactants: Molecular intercations, assmbly, and applications. Surface Science Reports. 2017;**72**(1):1-58

[229] Kang H, Buchman JT, Rodriguez RS, Ring HL, He JY, Bantz KC, et al. Stabilization of silver and gold nanoparticles: Preservation and improvement of plasmonic functionalities. Chemical Reviews. 2019;**119**(1):664-699

[230] Zhang T, Song YJ, Zhang XY, Wu JY. Synthesis of silver nanostructures by multistep methods. Sensors. 2014;**14**(4):5860-5889

[231] Chen MJ, He YR, Liu X, Zhu JQ, Liu R. Synthesis and optical properties of size-controlled gold nanoparticles. Powder Technology. 2017;**311**:25-33

[232] Bastus NG, Comenge J, Puntes V. Kinetically controlled seeded growth synthesis of citrate-stabilized gold nanoparticles of up to 200 nm: Size focusing versus Ostwald ripening. Langmuir. 2011;**27**(17):11098-11105

[233] Wu KJ, Torrente-Murciano L. Continuous synthesis of tuneable sized silver nanoparticles via a tandem seed-mediated method in coiled flow inverter reactors. Reaction Chemistry & Engineering. 2018;**3**(3):267-276

Chapter 4

Rice Husk Nanosilica Preparation and Its Potential Application as Nanofluids

Huei Ruey Ong, Wan Mohd Eqhwan Iskandar and Md. Maksudur Rahman Khan

Abstract

The fast development in the extraction technique of silica from biomass has resulted in the signification use of silica in the industry. Rice is one of the world's most significant plants, which serve as a carbohydrate intake for humans. Rice husk is one of the main agro-wastes comprising big quantities of silicate. This chapter presenting the review on rice husk nanosilica production techniques by thermal and chemical methods. A direction on efficient and sustainable nanosilica extraction method will be discussed. Apart from that, method on nanofluids preparation will be accumulated with respect to the end application. Moreover, the influence of nanoparticle in nanofluids in terms of heat conductivity, rheological properties, and stability will be discussed. The potential application area of silica nanofluids such as solar, automobile, electronic cooling, and biomedical application will be explored.

Keywords: rice husk nanosilica, chemical method, thermal method, industrial application

1. Introduction

The surface area and porosity of the nanosilica are large and can be commonly used in products such as fillers [1], pharmaceuticals [2], catalysts [3], and chromatography [4]. Industrial silica production uses sodium silicate as the main ingredient of silicone. Nevertheless, a large amount of energy is required to produce sodium silicate via melting the quartz sand and sodium carbonate at 1300°C [5]. In the future, fossil fuel energy may not be viable. Thus, it is also fascinating to create a technique for producing nanosilica from a silicon-containing biomass content that will be economically feasible. Biomass is a significant resource for renewable energy and represents 15% of the worldwide power supply [6]. Rice husk (RH) is one type of biomass, which is effective heat deliver and lignocellulose rich for biological oils [7]. The global annual product of RH is about 100 million tons [8]. RH is rich in silica content (~20 wt%) and abundant in rice milling as waste. RH is not widely known due to lack of commercial utilization. Nanosilica precursor is an exciting future application for the preparation of advanced materials, such as carbon/silica composites [9], photocatalysts [10], hydrogen production as well as CO_2 capture materials [11, 12], and metal ion removal adsorbents [13]. Nanosilica with porous RH composition can be prepared by various methods [14–16].

Kalapathy et al. [17] explored sodium hydroxide dissolved xerogel formation utilizing RH as raw resources. They discovered that combining the rice husk ash (RHA) acid with xerogel's washing step can efficiently improve nanosilica sample purity. Following a pre-treatment with acid, Zhang et al. [18] utilized RH as a forerunner to acquire superfine 30–200 nm diameter nanosilica the pretreated sample. In the latest studies, biotransformed nanosilica with *Fusarium oxysporum* fungus [19] or via a bio-digestion process using worms [20]. Witoon et al. [21] utilized RHA as raw resources for the preparation of bimodal porous nanosilica and Chitosan as a template.

Meanwhile, nanofluid is comprised of nanometer-sized particles (nanoparticles) and fluids. Water, engine oil, ethylene glycol, and so on are usually used for base fluids in many industries including transport, power supply, manufacturing, and electronics [22]. Conventional base fluids suffer from low heat transfer performance, which limits its application [23]. In order to overcome the drawbacks, nanosized particles suspended in the base fluid can improve the transfer of heat and rheological properties, acting as property enhancer [24]. Moreover, most of the nanofluid studies underline the nanoparticle preparation methodology. A research from Rao et al. [25] found that nanofluids have greater thermal conductivity than conventional fluids, strongly nonlinear temperature dependence on effective thermal conductivity, improve or decrease heat transfer in single-phase flow, improve or decrease nucleate pool boiling heat transfer, and yield higher critical heat flows under pool boiling conditions. To the best of our knowledge, RH-derived nanosilica has not been reported elsewhere. In this context, the method of preparing nanosilica will be deliberated. Moreover, the method of nanofluid preparation from nanoparticle and the potential applications of nanofluids will be discussed.

2. Method of preparing nanosilica

Thermal and chemical methods are the two major methods that have been widely adopted for silica production from biomass. **Figure 1** illustrated the methods used for producing nanosilica from biomass/agricultural waste.

2.1 Thermal methods

Thermal methods involve the utilization of furnace muffles, fixed bed furnace, fluidized bed reactor, and other thermal methods that consist of inclined step-grate furnace, cyclone furnace, and rotary kiln. The thermal technology does have a number of disadvantages such as required more time for reaction, hot spot formation, the absence of free-flowing air for full carbon oxidation, and many others [26].

2.1.1 Electric/muffle furnace

Nanosilica is extracted from agricultural waste in a laboratory scale by electric/muffle furnace. The biggest disadvantage in using this technology is the long reaction time and a lower production rate. Patil et al. [27] investigated the biggest RH nanosilica extraction, consisting in thermal treatment with electric oven for 6 hours at 700°C at different temperatures. XRD and FTIR were used to characterize the sample. XRD information showed that the nanosilica acquired was amorphous in nature. About 95.55% pure nanosilica obtained from RHA with acid leaching preceded by the treatment of thermal heating with muffle oven at 600°C [28]. According to Bogeshwaran et al. [29], silica extracted from RH is highly

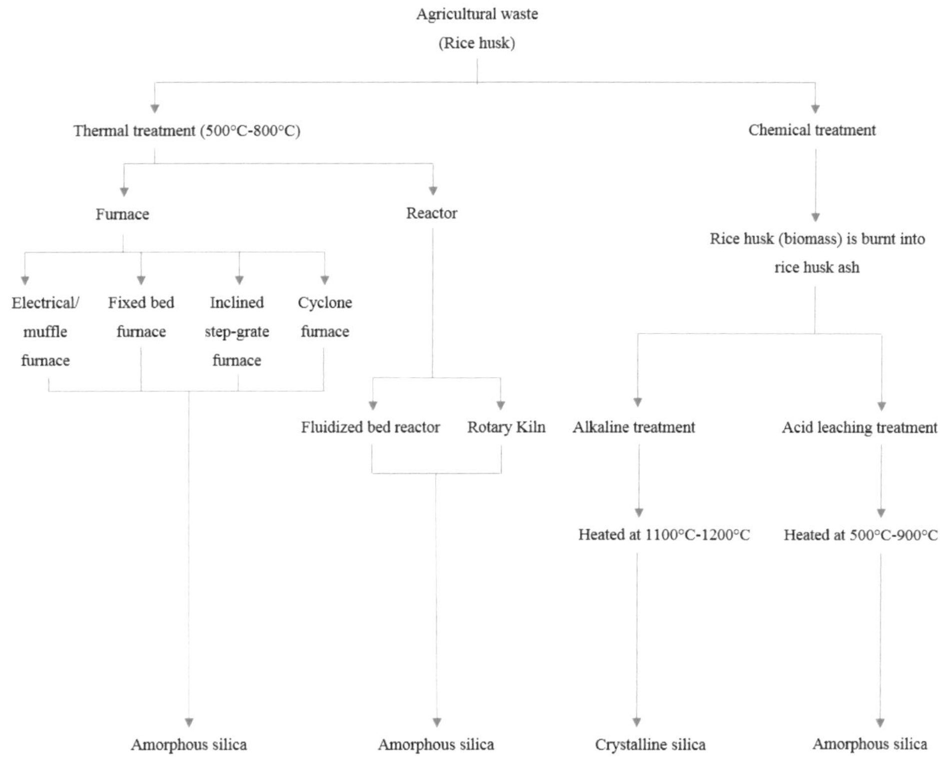

Figure 1.
Various treatments used to produce nanosilica from agricultural waste.

pozzolanic when burned in the muffle furnace. By thermal treatment, Chen et al. [30] utilized wheat straw to effectively produce nanosilica products. The combustion of wheat straw ash was kept at 500°C for 8 hours. The collected sample was washed with distilled water after the combustion and followed by calcination at the temperature from 400 to 700°C in a muffle furnace. Nano-amorphous silica was characterized by using XRD, TEM, EDX, FTIR, and BET. Ahmad Alyosef et al. [31] investigated the use of thermo-chemical treatment for meso/macroporous biogenic silica (3–1500 nm) from biomass such as miscanthus, wheat straw, and cereal remnant pellets. The biomass (wheat straw) was leached by concentrated H_2SO_4 (5 M). The wheat straw proportion of H_2SO_4 was controlled at 1:10 (gmL^{-1}). The treatment was performed under continuous stirring (1000 rpm) at 353 K for 24 hours. The ash of silica generated at various temperatures and times after heating by furnace. The combustion and acid leaching therapy of RH obtained pure amorphous silica. HCl, H_2SO_4, and HNO_3 leached the husk with different concentrations. The wheat-husk ash samples were positioned inside the muffle furnace at the temperature from 300 to 700°C for 24 hour after leaching treatment. The research proves that hydrochloric acid leaching treatment was more effective than any other acid to remove metal ions. Pure amorphous silica from acid-treated wheat husk ash was obtained at 500–700°C [32]. Yalcin and Sevinc [33] manufactured amorphous silica RH successfully at 600°C in a tubular stainless steel reactor for 4 hours in an electronic laboratory muffle furnace. In particular, electric/muffle furnace can increase the purity of silica contents obtained from incineration. Except that, amorphous silica structure can be obtained by incineration up to 425°C for 90 minutes. The structure of silica varies on the incineration temperature and time required [34].

2.1.2 Fixed bed furnace

The manufacturing of RH silica was also carried out using a fixed bed furnace. By using fixed bed furnace, Yang et al. [35] obtained amorphous silica in burst nano size. In this process, RH treated with raw and acid was conducted in fixed bed furnace for pyrolysis at 600–1200°C. The amorphous silica transforms into crystalline at 1000°C. Hamad [36] discovered RHA silica successfully using the 500–1150°C muffle furnace and fixed bed reactor.

2.1.3 Fluidized bed reactor

The advantages of fluidized bed reactor are the distribution of uniform temperature, fast reaction time, efficiency of carbon conversion, low temperature operating range, high intensity of combustion, elevated reaction of gas-solid mixtures, and outstanding mixing characteristics [37, 38]. Huang et al. [39] manufactured RH silica white by utilizing fluidized bed reactor. RH amorphous silica can be obtained by using fluidized bed bubbling pilot plants at different temperatures and at different speeds [40]. Genieva et al. [41] obtained RH silica material that is produced by the rice-milling phase, and it is a large agricultural waste product by using and characterizing the fluidized bed reactor throughout the nitrogen atmosphere. Luan and Chou [42] found RH silica in a modified fluidized bed reactor throughout the existence of pilot flame. Therefore, outcome revealed that the high-activity silica product was acquired.

2.1.4 Other thermal method

Inclined step-grate furnace is commonly used in the manufacturing of RHA. It consists of feeding component, chamber of combustion, and chamber of ash precipitation. The disadvantage of using this RHA manufacturing methodology is low yield quality and elevated unburnt carbon content. RH was provided from the upper part of the reactor as air flows from the lower part [43]. Moreover, cyclonic furnace was developed by Singh et al. [44]. In this furnace, the air kept the husk spinning and accelerated the combustion in the chamber. The benefit of using cyclone furnace to make husk ash is that the product has less carbon content. Subsequently, rotating kiln is a pyro-processing tool used in the ongoing process to increase calcination materials. Sugita [45] patented active RHA generated from rotary kiln. In this process, RH has been carbonized by an upstream rotary kiln that is heated at 300–400°C by electric heaters, burners, or other heat sources. Carbonized RH is supplied into rotating oven and burnt at 600°C after carbonization. These techniques effectively produced the husk ash. The disadvantage of using this technique was the need for additional fuel to avoid ash from being crystallized, longer reaction time, and high energy required.

2.1.5 Discussion on thermal method

Thermal method is one of the initial initiatives to obtain silica nanoparticle derived from RH biomass (**Table 1**). Muffle furnace helps in incineration of RHA to form nanosilica. The crystalline of nanosilica is dependent on the temperature and before incineration process takes place. Utilizing temperature around 500–700°C will form amorphous nanosilica. Alternatively, crystalline structure of nanosilica obtains above temperature of 900°C [46]. Chemical pre-treatment is vital to avoid any unburned material that leads to reduce the nanosilica's purity. Fixed bed furnace has an ideal temperature of 600–700°C to obtain white RHA. Complete combustion

DOI: http://dx.doi.org/10.5772/intechopen.89904

No	Method	Material	Time	Temperature (after getting silica gel)	Size	Purity	Yield	References
Electric/muffle furnace								
1	Furnace in which the subject material is isolated from the fuel and all of the products of combustion, including gases and flying ash	RH	700°C	6 hours	—	95.55%	—	Patil et al. [27]
2		RH	500°C	8 hours	—	—	—	Chen et al. [30]
3		RH	80	24 hours	3–1500 nm			Ahmad Alyosef et al. [31]
Fixed bed furnace								
1	Material is heaped onto a grate, and preheated primary air (called under fire air) is blown from under the bed to burn the fixed carbon	RH	600–1200°C	—	1–10 nm	—	—	Yang et al. [35]
		RH	500–1150°C	—	—	—	—	Hamad [36]
Fluidized bed reactor								
1	A simple fluidized reactor consists of a room, which is assisted by a distributor plate, and contains a bed of inert particles like sand	RH	800–950°C	4–8 hours	—	92–96%	—	Pitt [37] and Soltani et al. [38]
2		RH	100°C	4 hours	20 nm	—	—	Genieva et al. [41]
3		RH	60–860°C	4 hours	—	—	—	Luan and Chou [42]

Table 1.
List of thermal method and its parameters.

of carbon content is the major benefit of this furnace. However, the heat loss during the process could affect the temperature, which leads to unstable production of silica structure [36]. Fluidized bed reactor has many benefits such as high combustion intensity, lower operating temperature range, simple operation and quick start-up, and easier ash removal. However, it appears hard to fluidize RH and husk char or otherwise blended with sand, mold, and ash to produce a multi-structure [47–49]. Inclined step grate furnace is simple in construction and process, but it is inefficient in combustion and separation of ash resulted smoke and spark partially drawing into the dryer plenum [43]. The rotary kiln carbonizes RH first by burning without flaming and transforms the carbonized RH into ash. This method easily produces white RHA, which has excellent chemical reactivity [45]. This furnace requires new improvement to the capacity part due to low production along the process. Soponronnarit et al. [50] prove that cyclone furnace able to increase the furnace efficiency by 16% rises the air by 90%. Observation made proves that the height of ash on the grate does not affect the efficiency of the furnace. However, incomplete combustion may occur because of too high airflow rate in tertiary duct that did not support combustion since the burning RH fell quickly from the grate. Among them, fluidized bed reactor suites the best requirement for producing silica due to its better purity (92–96%) and operating at optimum temperature (800–950°C), which is also in agreement with Soltani et al. [38].

2.2 Chemical method

Chemical techniques include techniques of alkaline extraction used to achieve pure and high silica quantities. However, this method is costly due to a slightly longer reaction time (24–48 hours) and involves different measures with the use of different sorts of chemicals. Usually, RH will go through thermal process (incineration) to obtain RHA before proceeding to any chemical process involvement.

2.2.1 Alkaline extraction

Alkaline extraction and acid neutralization are an effective and easy technique of extracting amorphous silica from agricultural waste. Zulkifli et al. [51] utilized alkali extraction technique from RH to extract silica particles in order to remove metallic impurities. In a water bath, RHA was initially treated with HCl for 4 hours at 75°C. The filtration took place by constantly washing using distilled water until neutral state was reached and dried at 110°C for 12 hours. The NaOH was used to prepare a constantly stirring solution of sodium silicate for 1 hours at 90°C. The silicate sodium solution was then reacted to ethanol, and a steady 10-minute water mix was added. The whole mixture has been titrated 3 M H_3PO_4 until gel formation is carried out. The product after centrifugation of yellowish gel was washed with distilled water to clear away residual sodium silicate and phosphate, followed by calcination to produce silica nanoparticles. Hassan et al. [52] prepared nanosilica from rice husk in high surface area using the NaOH (alkaline extraction method). In their study, analyses of characterization of nanosilica were investigated by using FTIR, XRD, SEM, and TEM. The impact from their study states that more than 95% of nanosilica obtained. Liou and Yang [53] investigated various variables of silica derived from RHA processing via the alkali-extracted method. Acid and alkaline concentrations, gelation pH, aging time, and temperature have been optimized to prepare SiO_2 nanoparticles from RHA. The effects on the surface area from various acids and silica particle size have also been assessed. Rehman et al. [54] synthesized nanosilica using silica source from RHA. Silica nanoparticles were obtained from RH through the use of NaOH alkaline sol-gel method.

The application of H_2SO_4/water/butanol to pH 4 precipitated the silica. Thuc and Thuc's [55] technique was used to obtain nanosilica particles with high-specific surface area. Their study continues to prepare zeolite Y in sodium form (NaY) derived from nanosilica. Awizar et al. [56] produced and used nanosilica as a green corrosion inhibitor by alkaline extraction. Haq et al. [57] obtained RHA silica with reflux condition for a varying period of time by NaOH solution. The RHA reaction mechanism mixed with NaOH was given as follows:

SiO_2	+	2 NaOH	→	Na_2SiO_3	+	H_2O
(white ash)		(sodium hydroxide)		(sodium silicate)		(water)

Silica was precipitated by sodium silicate acid neutralization [55].

Na_2SiO_3	+	HCl	→	SiO_2	+	NaCl	+	H_2O
(sodium silicate)		(hydrochloric acid)		(silica gel form)		(sodium chloride)		(water)

Low surface microsphere silica can be achieved by alkaline and acid precipitation from wheat husk ash. Nano amorphous silica with a specific surface area of 8.23 m^2/g was achieved after alkaline extraction with NaOH [58]. Masnar and Coorey [59] prepared silica nanoparticles by following the same step as Liou and Yang [53]. Silica nanoparticles obtained at 80°C for 48 hours after solids have been dried.

Selvakumar et al. [60] prepared silica from RHA by adopting pre-treatment process (acid process). Pre-treatment of acid was used to enhance the silica purity with the effective removal of the majority of metallic impurities and to produce silica (white color). From their study, RHA was pre-treatment by various acids (pH 1, 3, 5, or 7 using 6 N hydrochloric acid, nitric acid, and sulfuric acid). RHA amorphous nanostructured silica was produced using alkaline extraction technique with NaOH solutions (2.0–3.0 N). Their research showed that treatment with 2.5 N NaOH produced RHA containing 90.44% silica. Rungrodnimitchai, Phokhanusai, and Sungkhaho [61] prepared RHA silicate materials using 2.0 M of sodium hydroxide with the help of microwave (800 W) for 10 minutes. Zhang et al. [62] synthesized silica nanoparticles from RHA by involving acid pre-treatment. Na_2CO_3 solution was added after the pre-treatment to obtain nanosilica slurry. The precipitation was then cleaned with distilled water and dried for 24 hours in the vacuum oven at 120°C. Adam et al. [63] obtained spherical nano size silica from RH by using nitric acid (65%) and sodium hydroxide. No calcination for ash formation was required in this treatment.

2.2.2 Acid extraction

Faizul et al. [64] prepared amorphous nanosilica with the size of 181.2 nm with mild acid solution (citric acid, acetic acid, and phosphoric acid) obtained from rice husk. Carmona et al. [65] used acid leaching to synthesize nanosilica of two kinds of rice husk, namely the agulhinha and the catetus. They believe that their method can be efficient in removing impurities (Zn, MN, Ca, K, Mg, Cu, and Al). Mahmud et al. [66] used hydrochloric acid for acid leaching to obtain high purity and high surface area of nanosilica. Rafiee and Shahebrahimi [67] prepared nanosilica from rice husk with high surface area by acid leaching treatment. The average size of nanosilica is 6–7 nm supported by the catalyst. Bakar et al. [68] prepared high purity silica by acid treatment followed by combustion. Pre-combustion rice husk was leached with hydrochloric acid and sulfuric acid to achieve pure silica. Thus, XRF confirmed the purity of amorphous silica over 99%.

2.2.3 Other chemical extraction methods of silica from agricultural waste

Many chemical treatments exist for the production of silica from bio-waste. Faizul et al. [69] obtained amorphous silica and activated carbon by three effective procedures by using toluene/ethanol, $NaClO_2$, and KOH. The method of calcination was used in the production of nano amorphous silica (100–120 nm). The manufacturing of amorphous silica was carried out using organic acid leaching instead of strong acid [70]. Ionic liquid was also used in the manufacturing of silica from agricultural waste by Kumar et al. [71].

2.2.4 Discussion on chemical method

Chemical method is advisable to obtain high purity of amorphous nanosilica due to its effective chemical reaction (**Tables 2** and **3**). Basically, there are two types of extraction methods (alkaline extraction and acid extraction). In this review, alkaline extraction method is predominantly compared to acid extraction method in terms of nanosilica properties obtained. Hassan et al. [52] produced the preparation of silica nanoparticle by alkali treatment and obtained more than 95% purity of nanosilica. Furthermore, Liou and Yang [53] prepared nanosilica and obtained 99.48% of silica content throughout alkali treatment. For further improvement, Selvakumar et al. [60] used pre-treatment and result in high purity (85%) of silica contents. Similarly, Adam et al. [63] also reported acid pre-treatment before conventional alkaline method, where ~95% purity of nanosilica was obtained. As for pre-treatment is use to enhance silica purity and remove metallic impurities. Meanwhile, Rungrodnimitchai et al. [61] used 2.0 M sodium hydroxide assisted by microwave (800 W) to obtain high purity of nanosilica from RHA. The modification could enhance the properties of the nanosilica obtained from the conventional method in terms of morphology, size, and purities as presented in **Table 2**. Furthermore, acid leaching method has been presented in **Table 3**. It was found that acid leaching method produced high purity of nanosilica as reported by Bakar et al. [68], where 99% purity of nanosilica was obtained with 500–700 nm. Similarly, Mahmud et al. [66] also reported that acid leaching method produced 99% high purity of nanosilica using HCl. Referring to above, acid leaching improves the other metal removal and increases the purity of nanosilica. It is noticed that single method like alkaline extraction and acid leaching method required high temperature thermal process to acquire nanosized silica. While combination of acid leaching and alkaline method could provide high purity of nanosilica without high temperature thermal process, in another words, mild condition, as reported by Adam et al. [63] and Selvakumar et al. [60].

2.2.5 Discussion for method

As mentioned above, two major methods that have been widely adopted by researcher in nanosilica production are thermal and chemical methods. Fluidized bed reactor could produce high purity of nanosilica at 92–96% using thermal process at 800–950°C for 4–8 h [38]. While chemical method modified alkaline method showed promising properties, produced 95% purity nanosilica with 110°C (mild condition) [63]. It is noticed that thermal method used a lot of energy (high temperature and long reaction time) to acquire nanosilica, whereas chemical method required high usage of chemicals (acid and alkaline solution), which

No	Method	Material	Time	Temperature (after getting silica gel)	Size	Purity	Yield	References
Alkaline extraction method								
1	Using alkali solution and followed by acid neutralization and undergo thermal process (calcination)	RH	30 minutes	550°C	98–272 nm	—	—	Zulkifli et al. [51]
2		RH	4 hours	700°C	20–25 nm	>95%	—	Hassan et al. [52]
3		RH	48 hours	80°C	20–30 nm	99.48%	91.91% @pH 3	Liou and Yang [53]
4		RH	24 hours	50°C	10–20 nm	—	—	Awizar et al. [56]
6		RHA	24 hours	60°C	—	—	80% (NaOH concentration of 1.0 mol dm^{-3})	Haq et al. [57]
7		Wheat husk ash	1 hour	550°C	227 nm	—	—	Cui et al. [58]
8		RHA	48 hours	80°C	—	50.15%	—	Masnar and Coorey [59]
Alkaline modification method								
1	Acid pre-treatment before conventional alkaline extraction	RHA	1 hour	130°C	—	85% (1.0 N NaOH)	—	Selvakumar et al. [60]
3		RHA	24 hours	120°C	47 nm		69–73% (250 ml Na_2CO_3 solution)	Zhang et al. [62]
5		RH	overnight	110°C	15–91 nm	~95.5%	—	Adam et al. [63]
2	Separating rice husk ash silica gel from microwave heating	RHA	48 hours	150°C	—	—	—	Rungrodnimitchai et al. [61]

Table 2.
List of alkaline technique based on its parameters.

Acid leaching method								
No	Method	Material	Time (h)	Temperature (after getting silica gel)	Size	Purity	Yield	References
1	Raw material undergoes acid leaching at mild condition and followed by thermal process (calcination)	Palm ash	30 minutes	800°C	—	92% (6% citric acid)	—	Faizul et al. [64]
2		RH	1 hour	650°C	181.2–294.7 nm	—	—	Carmona et al. [65]
3		RH	2 hour	700°C	53–55 nm	99.761% (HCl), 99.760% (citric acid)	—	Mahmud et al. [66]
4		RH	48 hours	50°C	6 nm	98.801%	—	Rafiee and Shahebrahimi [67]
5		RH	2 hours	600°C	500–700 nm	>99%	—	Bakar et al. [68]

Table 3.
List of acid leaching method based on its parameters.

resulted in cost intensive. Moreover, high thermal and chemical methods also contributed some bad impact on environment due to releasing of nonfriendly gases and waste materials produced, respectively. Thus, low cost and environmental friendly method is required to idealize for industrial application. In recent year, Mor et al. [72] reported a low-cost method in preparing nanosilica using green technology. Initially, the RHA was dissolved in NaOH and placed to autoclave at 100°C for 2 h to obtain the mixture slurry and followed dilution with distilled water for phase separation. The supernatant proceeds for silica extraction with filtration process. The filtrate precipitated with HCl and followed by washing and oven dried at 50°C where high purity of nanosilica (99%) was obtained.

3. Method of nanofluid preparation

There are two main methods for preparing nanofluids, which are one-step and two-step methods. One-step method combines between synthesis and dispersion of nanoparticles into base fluid in one step. Several differences exist in these methods. In one of the conventional techniques called the one-step method of direct evaporation, the nanofluid is obtained inside the base fluid by solidifying the nanoparticles that are originally in the gaseous phase. Akoh et al. [73] created the one-step direct evaporation method and are referred to vacuum evaporation on the method of running oil substrates. The concept of this method was originally developed the nanoparticles in order to obtain dry nanoparticles. Particles are difficult to differentiate from liquids. The technique of laser ablation to obtain alumina nanofluids is another one-step technique [74]. Zhu et al. [75] used one-step technique to prepare copper nanoparticles in the medium of ethylene glycol.

The two-step method is widely utilized for nanofluid preparation, and most of the cases used nano powders (solid) during the preparation. The technique first produces nanoparticles; thereafter, the nanoparticles will be dispersed into the base fluids. Jena et al. [76] used hydrogen reduction techniques to prepare nanoparticles from the chemical precursor and dispersed them into fluid via two-step methods. The use of ultrasonic technique to disperse the nanoparticles into deionized water, which containing sodium lauryl sulfate (SLS) during nanofluid preparation, is also one of the widely adopted technique [77].

Wei and Wang [78] synthesized copper nanofluids by using a constant flow microfluidic microreactor. Through this technique, the microstructure copper nanofluids can be synthesized continually by changing parameters such as additive and flow rate and reactant concentration. Using a new precursor conversion technique, ultrasonic and microwave irradiation can be used to synthesize CuO nanofluids with a better solid volume fraction (~10 vol%) [79]. Under microwave irradiation, the $Cu(OH)_2$ precursor will entirely converted into CuO nanoparticles in H_2O. The ammonium citrate stops nanoparticles from growing and aggregating, resulted in stable CuO nanofluid with a better heat conductivity than the ones produced by using other dispersive techniques. The easier way to acquire colloids of monodisperse noble metal is by using the technique of phase transfer [80]. The two-phase cyclohexane system, aqueous formaldehyde, is transmitted to cyclohexane in water through the dodecyl amine response to cyclohexane as an intermediate form reduction. Cyclohexane solution intermediates can reduce Ag or Au ions in aqueous solution to form dodecyl amine-protected Ag and Au nanoparticles at room temperature. Feng et al. [81] used phase transfer method in preparing Au, Ag, and Pt nanoparticles based on a reduction in solubility of PVP in water at increased temperature. The technique of phase transfer is also used to prepare stable Fe_3O_4 nanofluids based on kerosene. Oleic acid is effectively grafted in chemisorbed fashion on the surface of Fe_3O_4 nanoparticles, enabling Fe_3O_4 nanoparticles to be well compatible with kerosene [82]. The phase-transfer technique prepared Fe_3O_4 nanofluids not showing "time reliance of the characteristic of heat conductivity" as reported previously. The main problem is the production of nanofluids with a controllable microstructure. It is well recognized that nanofluid characteristics are highly dependent on nanomaterial structure and shape. Recent study demonstrates the improvement in conductivity and the stability of nanofluids when synthesized using chemical solution compared to other techniques [83]. This technique is differentiated by its controllability from others. The microstructure of nanofluids can be differed and manipulated by regulating the factors of synthesis, including acidity, radiation from the microwave and ultrasonic, temperature, acidity, concentrations, and types of reactor and the order of additives added to the solution.

Silica is widely used as both precursor and material for ceramic product manufactures. Silica has high abrasion resistance, high thermal stability, and electrical insulation [84]. Fazeli et al. [85] dispersed nanosilica into the distilled water, and the suspension was sonicated for at least 90 min in an ultrasonic bath. They discovered that silica nanofluids remained stable without visible settlement for 72 hours. Pang et al. [86] used ultrasonic to mix SiO_2-pure methanol by ultrasonic (750 W, 20 kHz) and Al_2O_3-pure methanol to break the agglomeration through vibration during 2 hours. They examined the impact of the zeta potential and pH of methanol-based nanofluids in nanoparticles. They proved that the zeta potential is closely connected with the pH of the suspension. Al_2O_3 nanofluids have zeta potential >60 mV; meanwhile, SiO_2 nanofluids have zeta potential >30 mV, which indicated that both nanofluids were well stable. The visualization and Tyndall effect (light dispersion study in nanoparticles) images show that nanofluids based on methanol are well dispersed. Bolukbasi and Ciloglu [87] have been using magnetic stirrer to prepare SiO_2 nanofluids. The suspensions were continuously sonicated for 2 hours into an ultrasonic vibrator (600 W and 40 kHz). The researcher confirmed that no sedimentation was traced throughout the experimental period. Darzi et al. [88] applied distilled water to the specified quantity of SiO_2 nanoparticles and mixed for half an hour with a magnetic stirrer. Afterward, the ultrasonic vibrator was dispersed for 2 hours to have the stable suspension. During the synthesis method, no surfactant/dispersant additives were added, otherwise affecting the thermophysical characteristics of nanofluid. Silica nanoparticles were used to function through grafting silanes directly on the silica nanoparticles surfaced by Yang and Liu [89]. For the functioning method, silane of (3-glycidoxylproyl) trimethyoxysilane has been used as the reacting silane and silica nanoparticles with a mass ratio of 0.115. Nanoparticles were successfully dispersed into water. Meanwhile, the solution was stored at 50°C for 12 hours of ambient temperature. Functional nanoparticles were discovered to continue to disperse even when the nanofluid remained at a mass concentration of 10% for 12 months. In addition, no sedimentation has been reported. They prepared traditional nanoparticles by dispersing and oscillating them to water. Powder of silica nanoparticles was first dissipated into deionized water and then oscillated in an ultrasound bath for 12 hours. Sedimentation was found after a few days. Anoop et al. [90] dispersed SiO_2 nanoparticles with an ultrasound bath in deionized water for 30 minutes. In addition, the application of a sonicator type probe to the nanofluids intensified this colloidal suspension. The suspension was provided by cyclic ultrasonic pulses for around 15 minutes in order to obtain maximum particle de-agglomeration. By adding nitric acid reagent grade from the isoelectric pH value, the pH value of the suspension was kept at 4.5. Nanofluids have been indicated to show excellent stability over period. Qu and Wu [91] developed nanofluids Al_2O_3 and SiO_2 water. The pH value of the nanofluids was modified as a first step to a value that was far from the respective isoelectric point

(IEP) of silica (with pH ~ 3) or alumina (with pH ~ 9), then added to the water nanoparticles (with pH ~ 9) and with pH ~9. The dispersion solution was vibrated in an ultrasonic bath for about 4 hours afterward. Alumina nanoparticles have been discovered to be better dispersed. Hwang et al. [92] generated CuO, MWCNT, and SiO_2 nanofluids by using an ultrasonic disruptor. For SiO_2 and CuO nanoparticles, they acquired stable suspensions. Nevertheless, sodium dodecyl sulfate (SDS) has been used as a surfactant to produce MWCNT nanofluids as the MWCNTs are entangled and aggregated into aqueous suspension.

4. Potential application of nanofluids

Nanofluids have been proved in experiment and theory in enhancing heat transport and energy efficiency for various manufacturing purposes such as mechanical applications, electronic cooling, transportation, and many more in a range of thermal exchange technologies. In all applications, nanofluid performs a key position in creating the next device generation for various medical and engineering applications. Some of the following applications are discussed below.

4.1 Solar applications

The temporal difference between energy supply and energy requires rendered storage system design. Stocking of thermal electricity as in solar thermal installations as sensitive and latent heat, with an emphasis on an effective use as well as preservation of wastewater and solar energy in buildings and manufacturing, has become an significant element in energy planning [93]. Compared to the basic material, the PCMs contained extremely high thermal conductivity. Liu et al. [94] synthesized a new type of nanofluid phase change material (PCM) with a tiny portion of TiO_2 nanoparticles suspended in aqueous saturated $BaCl_2$ solution. The PCM nanofluids had relatively better thermal conductivity compared to base material. The cool storage/supply rate and the cool storage/supply capability have risen significantly compared with aqueous solution of $BaCl_2$ without the need of additional nanoparticles. The greater thermal characteristics of PCMs show that in cool storage applications, they have the ability to replace standard PCMs. Copper nanoparticles are the additives that are efficient to enhance PCM cooling and heating levels. Shin and Banerjee [95] recorded an anomalous increase in nanofluid-specific heat capacity of high temperature. The researcher discovered that 1 wt% SiO_2 nanoparticle-doped alkali metal chloride salt eutectic improves the specific thermal capacity of nanofluid by ~15% to be used in solar thermal energy storage facilities. One of the methods used to store solar energy is the use of PCMs. Paraffin is the most appropriate of many accessible PCMs because of its attractive features, including large latent heat capacity, insignificant super cooling, and low cost. The intrinsic low thermal conductivity (0.21–0.24 W/mK), however, avoids possible applications [96]. Wu et al. [96] numerically researched Cu/paraffin nanofluid PCM melting procedures. Their findings showed that the melting time with 1 wt% Cu/paraffin is saved by 13.1%. The study found that the addition of nanoparticles is an effective way for increasing the heat transfer of latent heat energy storage system.

Solar energy is an important factor in energy use because of a shortage of electricity generation. Lack of fossil fuel and environmental factors will limit future use of fossil fuels. Researchers are encouraged to discover alternative energy sources. This became even more widespread as fossil fuel prices continue to increase. In latest years, solar energy has had a notable advantage. In just 1 hour, the earth gets more sun energy than the world consumes for a year [97, 98].

Solar collectors are specific types of heat exchangers that convert solar energy to transport medium internal energy. This equipment absorbs incoming sunlight, which is converted into heat and transmitted the heat to a fluid that flows through the collector (generally oil, air, and water). The energy is collected directly from the working fluid to the hot water or space conditioning or thermal energy storage tank, for night or on cloudy days [99].

Taylor et al. [100] found that the use of the graphite/therminol VP-1 nanofluid with volume fractions around 0.001% or less could be of benefit for 10–100 MWe energy crops. In combination with a solar thermal power tower with 100 MWe of capacity in a solar resource such as Tucson, Arizona, the researchers estimated that $3.5 million more could be achieved each year. The supply of fresh water is more crucial arid distant areas of the globe. Solar desalination technologies are possible to overcome portion of the issue in these areas, where solar energy is accessible. The absence and untrustworthy drinking water is a main issue in developing countries. Global dryness and desertification are estimated to make drinking water a major problem in the world [101].

Greenhouse gas emission from fresh water production can be prevented by solar stills [102]. Many experts have researched solar stills and used different techniques to enhance their productivity. Gnanadason et al. [103] found that the productivity of solar system was influenced by nanofluids. The implications of putting carbon nanotubes (CNTs) to the water in a single solar basin were investigated. The findings have shown the addition of nanofluids that will enhance the efficiency by 50%. However, the quantity of nanofluid added to the water inside the solar was not yet mentioned. In addition to solar nanofluids, the economic growth should be perceived. Certain works in the literature disclosed the addition of dyes to solar stills could increase the efficiency. Nijmeh et al. [104] investigated that adding violet color to the solar water still improves the efficiency significantly by 29%. Furthermore, nanofluids (especially the CNTs) are more expensive, and this might therefore be a difficult task for the use of nanofluids in solar stills because the nanofluids in solar stills do not flow in a closed loop in order to recover them.

4.2 Automobile applications

Adding nanotubes and nanoparticles to the conventional engine coolants (ethylene glycol and water mixture), nanofluid lubricants can boost their thermal conductivity and enhance heat change rates and fuel economy [105]. Tzeng et al. [106] have studied the impacts of nanofluids on automatic transmission cooling. They spread CuO and Al_2O_3 nanoparticles and antifoams into the transmission fluid and then used four-wheel automatic transmission on a real-time basis. The findings indicate that CuO nanofluid has the lowest temperature distribution and the highest heat transfer impact on the rotating speeds [107]. CuO and nanofluids based on aluminum oxides were developed with the arc-submerged nanoparticle synthesis system along with the plasma charging arc system [108, 109]. Both types of nanofluids have increased the characteristics, including a greater boiling point, a greater viscosity, and a greater conductivity than conventional brake fluid. With greater viscosity, conductivity, and boiling point, the brake oil nanofluids reduce the vapor lock from occurring and offer greater safety in driving condition [110].

4.3 Electronic cooling

As IC (embedded circuit) and microelectronic parts decrease in size, the energy dissipation has risen dramatically. Better thermal management and cooling liquids are necessary for secure operation, with enhanced heat transport characteristics.

Nanofluids were regarded as working liquids for electronic cooling applications in heat pipes. Tsai et al. [111] used water-based nanofluid as the operating channel for circular heat pipe. It was intended as a heat diffuser and applied in CPU of notebook or desktop PC. The findings exhibited that the nanofluid heat pipes have considerably lower thermal resistance than deionized water. The findings showed that the thermal strength of a vertical meshed heat tube differs respectively with nanoparticle size. Ma et al. [112] examined the impact of nanofluids toward oscillating heat pipe transport capability. The experimental results reveal that the temperature difference between the evaporator and the condenser decreased from 40.9 to 24.3°C at an input energy of 80 W by 1 vol% nanoparticles. Lin et al. [113] examined nanofluids using silver nanoparticles in heat pulsating pipes and found supportive outcomes. The silver nanofluid enhanced the thermal transfer properties of the heat pipes. Vafaei et al. [114] found that nanofluids are efficient in engineering surface wettability and potentially surface tension. With a goniometer, the presence of a very small bismuth telluride nanofluid concentration significantly affected the wetting features of the surface. Concentrations as low as 3×10^{-6} improved the contact angle to more than 40°, showing clearly nanoparticles affect the force balance triple line vicinity. Experimental, numerical, and theoretical studies on nanofluid prove numerous prospective applications of nanofluids are present such as electronic cooling, displays, micro devices, cameras, thermal exchangers, military, spacecraft equipment, boats, medicine, atomic reactors, fuel cell and sensor applications. The stability of nanofluids is a major challenge for nanofluid commercialization. By solving the problems, significant developments are anticipated in many applications. Further study should be conducted on numerous heat and fluid applications.

4.4 Biomedical applications

Some special types of nanoparticles possess antibacterial activity or drug delivery properties, so that nanofluids that contain these nanoparticles have certain relevant properties [110]. Organic antibacterial products, especially at high temperatures or pressures, are often less stable. Consequently, inorganic materials such as metal oxides and metal have received considerable attention in the previous decade because they are able to resist severe process circumstances. ZnO nanofluid antibacterial behavior indicates that ZnO nanofluids are bacteriostatic to *Escherichia coli*. With the growing concentration of nanoparticles, antibacterial activity rises and the particle size decreases. Measurements of electrochemical show a direct interaction between ZnO and elevated ZnO levels of bacterial membrane (L. [115]). Jalal et al. [116] created ZnO nanoparticles with a green technique. An estimation of the reduction ratio of ZnO-treated bacteria was made on ZnO suspension activity of nanoparticles with *E. coli*, the bacteria's survival ratio reduces with increased nanofluid ZnO levels and time. Silver nanoparticles were discovered to depend on the size of silver particles for their antibacterial activity. Antibacterial efficiency was achieved by the very small silver concentration of 1.69 mug/mL Ag [117]. Lyon and Alvarez [118] suggested that C60 suspensions exhibit ROS-independent oxidative stress in bacteria that show protein oxidation, modifications in cell membrane potential, and cellular respiration interruption. The mechanism needs direct contact between bacterial and nanoparticles as well as contrast from nanomaterial antibacterial processes earlier reported involving ROS generation (metal oxides), or leaching of toxic components (nanosilver).

4.5 Discussion for applications

It was found that nanoparticles added to the base fluid improve the characteristics of fluids such as structure, thermal conductivity, viscosity, convective heat transfer, density, and specific heat. In our review, we have narrowed down the application of nanofluids such as solar, automobile, electronic cooling, and biomedical application. It is noticed that the physical properties of nanoparticles such as size and crystallinity are influencing the nanofluid performance during its application. For instance, Micali et al. [119] explored the possibility to reduce temperature up to 13.6% on the exhaust valve seat and up to 4.1% on the exhaust valve spindle by 2.5% volume concentration on the cylinder head and the spindle of the exhaust valve. Al-Jethelah et al. [120] discovered improvements of solar thermal applications in terms of melting process through numerical and experimental by adding nanofluids into PCM. Said et al. [121] prepared 0.3% volume fraction of Al_2O_3 nanofluids and dispersed into distilled water and ethylene glycol as base fluid (ratio of 50:50) and discovered that it enhances the thermal performance by 24.21%. We believe that RH-derived nanosilica will provide similar performance compared to other semiconductor nanoparticles as mentioned above. Akilu et al. [122] attained ~27% thermal conductivity enrichment at 21.1% disparagement of specific heat by using hybrid nanofluids, and SiO_2-CuO (0.5–2 vol%) dispersed into base fluid (Glycerol/EG). Yao et al. [123] did the research on the boiling efficiency of Al_2O_3, SiO_2, and their mixture with water at the ratio of 1:1. The significance of their study was its impact of pressure on the performance of boiling nanofluids. Based on the outcomes, nanofluid efficiency increased the pressure reduction. Authors also regarded the effects of nanoparticle size on the heat flux posed tiny rise while raising the nanoparticle size between 30 and 50 nm.

5. Conclusion

This book chapter collectively reviews the preparation method of rice husk nanosilica, its application as nanofluids, and nanofluid application in the industry. There are two main methods in preparing nanosilica, namely thermal and chemical methods. It is noticed that chemical method is more preferable than thermal method in terms of nanosilica purity, which is critical. The popular chemical methods widely used by the researcher are alkaline extraction and acid leaching method. It has found that utilizing solely single chemical method must follow high thermal treatment and high operating cost, which is not feasible. Thus, a combination or modification of the chemical method is required to improve the purity of nanosilica. Pre-acid treatment followed by conventional alkaline extraction presented better purity of nanosilica. The purity of nanosilica is a crucial property in nanofluid preparation, which will affect the performance of the nanofluids. There are one-step and two-step methods, which are widely adopted by the researcher in preparing nanofluids. One-step technique combines the production of nanoparticles and dispersion of nanoparticles into the base fluid with a single step. Meanwhile, in two-step method, nanoparticles are first produced and then dispersed into the base fluids. However, two-step method is preferable for rice husk nanosilica-based nanofluid preparation, which involves ultrasonic method. The application of the nanofluids has been explored such as solar application, automobile application, electronic cooling application, and biomedical application. It was found that nanofluids could improve the base fluid performance due to the additional of the nanoparticles. Even though the review focused on semiconductor-based nanofluids, we believe that rice husk nanosilica-based nanofluids could also have the similar trends of performance. Gradually, the awareness on the usage of "green" material in the product is rising, and rice husk nanosilica could be an ideal candidate as nanoparticle and nanofluid application.

Author details

Huei Ruey Ong[1*], Wan Mohd Eqhwan Iskandar[1] and Md. Maksudur Rahman Khan[2]

1 Faculty of Engineering and Technology, DRB-HICOM University of Automotive Malaysia, Pahang, Malaysia

2 Faculty of Chemical and Natural Resources Engineering, Universiti Malaysia Pahang, Gambang, Pahang, Malaysia

*Address all correspondence to: roi_rui86@hotmail.com; hueiruey@dhu.edu.my

References

[1] Motaung T, Luyt A. Effect of maleic anhydride grafting and the presence of oxidized wax on the thermal and mechanical behaviour of LDPE/silica nanocomposites. Materials Science and Engineering: A. 2010;**527**(3):761-768

[2] Morpurgo M, Teoli D, Pignatto M, Attrezzi M, Spadaro F, Realdon N. The effect of Na_2CO_3, NaF and NH_4OH on the stability and release behavior of sol–gel derived silica xerogels embedded with bioactive compounds. Acta Biomaterialia. 2010;**6**(6):2246-2253

[3] Ge J, Huynh T, Hu Y, Yin Y. Hierarchical magnetite/silica nanoassemblies as magnetically recoverable catalyst–supports. Nano Letters. 2008;**8**(3):931-934

[4] Bakaev V, Pantano C. Inverse reaction chromatography. 2. Hydrogen/deuterium exchange with silanol groups on the surface of fumed silica. The Journal of Physical Chemistry C. 2009;**113**(31):13894-13898

[5] Affandi S, Setyawan H, Winardi S, Purwanto A, Balgis R. A facile method for production of high-purity silica xerogels from bagasse ash. Advanced Powder Technology. 2009;**20**(5):468-472

[6] Saxena R, Seal D, Kumar S, Goyal H. Thermo-chemical routes for hydrogen rich gas from biomass: A review. Renewable and Sustainable Energy Reviews. 2008;**12**(7):1909-1927

[7] Tang Z, Zhang Y, Guo Q. Catalytic hydrocracking of pyrolytic lignin to liquid fuel in supercritical ethanol. Industrial & Engineering Chemistry Research. 2010;**49**(5):2040-2046

[8] Yeletsky P, Yakovlev V, Mel'Gunov M, Parmon V. Synthesis of mesoporous carbons by leaching out natural silica templates of rice husk. Microporous and Mesoporous Materials. 2009;**121**(1-3):34-40

[9] Kumagai S, Sasaki J. Carbon/silica composite fabricated from rice husk by means of binderless hot-pressing. Bioresource Technology. 2009;**100**(13):3308-3315

[10] Artkla S, Kim W, Choi W, Wittayakun J. Highly enhanced photocatalytic degradation of tetramethylammonium on the hybrid catalyst of titania and MCM-41 obtained from rice husk silica. Applied Catalysis B: Environmental. 2009;**91**(1-2):157-164

[11] Bhagiyalakshmi M, Yun LJ, Anuradha R, Jang HT. Utilization of rice husk ash as silica source for the synthesis of mesoporous silicas and their application to CO_2 adsorption through TREN/TEPA grafting. Journal of Hazardous Materials. 2010;**175**(1-3):928-938

[12] Chen W-S, Chang F-W, Roselin LS, Ou T-C, Lai S-C. Partial oxidation of methanol over copper catalysts supported on rice husk ash. Journal of Molecular Catalysis A: Chemical. 2010a;**318**(1-2):36-43

[13] Wang L-H, Lin C-I. Adsorption of lead (II) ion from aqueous solution using rice hull ash. Industrial & Engineering Chemistry Research. 2008;**47**(14):4891-4897

[14] An D, Guo Y, Zhu Y, Wang Z. A green route to preparation of silica powders with rice husk ash and waste gas. Chemical Engineering Journal. 2010;**162**(2):509-514

[15] Pijarn N, Jaroenworaluck A, Sunsaneeyametha W, Stevens R. Synthesis and characterization of nanosized-silica gels formed under controlled conditions. Powder Technology. 2010;**203**(3):462-468

[16] Sun L, Gong K. Silicon-based materials from rice husks and their applications. Industrial & Engineering Chemistry Research. 2001;**40**(25):5861-5877

[17] Kalapathy U, Proctor A, Shultz J. A simple method for production of pure silica from rice hull ash. Bioresource Technology. 2000;**73**(3):257-262

[18] Zhang H, Zhao X, Ding X, Lei H, Chen X, An D, et al. A study on the consecutive preparation of d-xylose and pure superfine silica from rice husk. Bioresource Technology. 2010;**101**(4):1263-1267

[19] Bansal V, Poddar P, Ahmad A, Sastry M. Room-temperature biosynthesis of ferroelectric barium titanate nanoparticles. Journal of the American Chemical Society. 2006;**128**(36):11958-11963

[20] Estevez M, Vargas S, Castano V, Rodriguez R. Silica nano-particles produced by worms through a bio-digestion process of rice husk. Journal of Non-Crystalline Solids. 2009;**355**(14-15):844-850

[21] Witoon T, Chareonpanich M, Limtrakul J. Synthesis of bimodal porous silica from rice husk ash via sol–gel process using chitosan as template. Materials Letters. 2008;**62**(10-11):1476-1479

[22] Eastman JA, Choi S, Li S, Yu W, Thompson L. Anomalously increased effective thermal conductivities of ethylene glycol-based nanofluids containing copper nanoparticles. Applied Physics Letters. 2001;**78**(6):718-720

[23] Tajik Jamal-Abadi M, Zamzamian A. Optimization of thermal conductivity of Al_2O_3 nanofluid by using ANN and GRG methods. International Journal of Nanoscience and Nanotechnology. 2013;**9**(4):177-184

[24] Bahrami M, Akbari M, Karimipour A, Afrand M. An experimental study on rheological behavior of hybrid nanofluids made of iron and copper oxide in a binary mixture of water and ethylene glycol: Non-Newtonian behavior. Experimental Thermal and Fluid Science. 2016;**79**:231-237

[25] Rao KS, El-Hami K, Kodaki T, Matsushige K, Makino K. A novel method for synthesis of silica nanoparticles. Journal of Colloid and Interface Science. 2005;**289**(1):125-131

[26] Patel KG, Shettigar RR, Misra NM. Recent advance in silica production technologies from agricultural waste stream. Journal of Advanced Agricultural Technologies. 2017;**4**(3)

[27] Patil R, Dongre R, Meshram J. Preparation of silica powder from rice husk. Journal of Applied Chemistry. 2014;**27**:26-29

[28] Ghorbani F, Sanati AM, Maleki M. Production of silica nanoparticles from rice husk as agricultural waste by environmental friendly technique. Environmental Studies of Persian Gulf. 2015;**2**(1):56-65

[29] Bogeshwaran K, Kalaivani R, Ashraf S, Manikandan G, Prabhu GE. Production of silica from rice husk. International Journal of ChemTech Research. 2014;**6**:974-4290

[30] Chen H, Wang F, Zhang C, Shi Y, Jin G, Yuan S. Preparation of nano-silica materials: The concept from wheat straw. Journal of Non-Crystalline Solids. 2010;**356**(50-51):2781-2785

[31] Ahmad Alyosef H, Schneider D, Wassersleben S, Roggendorf H, Weiß M, Eilert A, et al. Meso/macroporous silica from miscanthus, cereal remnant pellets, and wheat straw. ACS Sustainable Chemistry & Engineering. 2015;3(9):2012-2021

[32] Chakraverty A, Mishra P, Banerjee H. Investigation of combustion of raw and acid-leached rice husk for production of pure amorphous white silica. Journal of Materials Science. 1988;**23**(1):21-24

[33] Yalcin N, Sevinc V. Studies on silica obtained from rice husk. Ceramics International. 2001;**27**(2):219-224

[34] Ramezanianpour A, Ahmadibeni G. The effect of rice husk ash on mechanical properties and durability of sustainable concretes. International Journal of Civil Engineering. 2009;**7**(2):83-91

[35] Yang H, Liu B, Chen Y, Li B, Chen H. Influence of inherent silicon and metals in rice husk on the char properties and associated silica structure. Energy & Fuels. 2015;**29**(11):7327-7334

[36] Hamad M. Combustion of rice hulls in a static bed. Energy in Agriculture. 1981;**1**:311-315

[37] Pitt N. Process for the Preparation of Siliceous Ashes: Google Patents; 1976; US3959007A

[38] Soltani N, Bahrami A, Pech-Canul M, González L. Review on the physicochemical treatments of rice husk for production of advanced materials. Chemical Engineering Journal. 2015;**264**:899-935

[39] Huang S, Jing S, Wang J, Wang Z, Jin Y. Silica white obtained from rice husk in a fluidized bed. Powder Technology. 2001;**117**(3):232-238

[40] Gomes GMF, Philipssen C, Bard EK, Dalla Zen L, de Souza G. Rice husk bubbling fluidized bed combustion for amorphous silica synthesis. Journal of Environmental Chemical Engineering. 2016;**4**(2):2278-2290

[41] Genieva S, Turmanova S, Dimitrova A, Vlaev L. Characterization of rice husks and the products of its thermal degradation in air or nitrogen atmosphere. Journal of Thermal Analysis and Calorimetry. 2008;**93**(2):387-396

[42] Luan TC, Chou TC. Recovery of silica from the gasification of rice husks/coal in the presence of a pilot flame in a modified fluidized bed. Industrial & Engineering Chemistry Research. 1990;**29**(9):1922-1927

[43] Bautista EU, Aldas RE, Gagelonia EC. Rice hull furnaces for paddy drying: The Philippine rice research institute's experience. ACIAR PROCEEDINGS; 1996:253-256

[44] Singh R, Maheshwari R, Ojha T. Development of a husk fired furnace. Journal of Agricultural Engineering Research. 1980;**25**(2):109-120

[45] Sugita S. Method of Producing Active Rice Husk Ash: Google Patents; 1994; US5329867A

[46] Chandrasekhar S, Pramada P, Majeed J. Effect of calcination temperature and heating rate on the optical properties and reactivity of rice husk ash. Journal of Materials Science. 2006;**41**(23):7926-7933

[47] Khan A, De Jong W, Jansens P, Spliethoff H. Biomass combustion in fluidized bed boilers: Potential problems and remedies. Fuel Processing Technology. 2009;**90**(1):21-50

[48] Natarajan E, Nordin A, Rao A. Overview of combustion and gasification of rice husk in fluidized bed reactors. Biomass and Bioenergy. 1998;**14**(5-6):533-546

[49] Rozainee M, Ngo S, Salema AA, Tan K. Fluidized bed combustion of rice husk to produce amorphous siliceous ash. Energy for Sustainable Development. 2008;**12**(1):33-42

[50] Soponronnarit S, Swasdisevi T, Wetchacama S, Shujinda A, Srisawat B. Cyclonic rice husk furnace and its application on paddy drying. International Energy Journal. 2007;**1**(2):67-75

[51] Zulkifli NSC, Ab Rahman I, Mohamad D, Husein A. A green sol–gel route for the synthesis of structurally controlled silica particles from rice husk for dental composite filler. Ceramics International. 2013;**39**(4):4559-4567

[52] Hassan A, Abdelghny A, Elhadidy H, Youssef A. Synthesis and characterization of high surface area nanosilica from rice husk ash by surfactant-free sol–gel method. Journal of Sol-Gel Science and Technology. 2014;**69**(3):465-472

[53] Liou T-H, Yang C-C. Synthesis and surface characteristics of nanosilica produced from alkali-extracted rice husk ash. Materials science and engineering: B. 2011;**176**(7):521-529

[54] Rehman MSU, Umer MA, Rashid N, Kim I, Han J-I. Sono-assisted sulfuric acid process for economical recovery of fermentable sugars and mesoporous pure silica from rice straw. Industrial Crops and Products. 2013;**49**:705-711

[55] Thuc CNH, Thuc HH. Synthesis of silica nanoparticles from Vietnamese rice husk by sol–gel method. Nanoscale Research Letters. 2013;**8**(1):58

[56] Awizar DA, Othman NK, Jalar A, Daud AR, Rahman IA, Al-Hardan N. Nanosilicate extraction from rice husk ash as green corrosion inhibitor. International Journal of Electrochemical Science. 2013;**8**(2):1759-1769

[57] Haq IU, Akhtar K, Malik A. Effect of experimental variables on the extraction of silica from the rice husk ash. Journal of the Chemical Society of Pakistan. 2014;**36**(3):382

[58] Cui J, Sun H, Luo Z, Sun J, Wen Z. Preparation of low surface area SiO_2 microsphere from wheat husk ash with a facile precipitation process. Materials Letters. 2015;**156**:42-45

[59] Masnar A, Coorey R. Application of sago pith waste and nanosilica from rice husk ash as hybrid bio-nanofiller composite for food plastic packaging. Ukrainian Food Journal. 2017;**6**(4):599-759

[60] Selvakumar K, Umesh A, Ezhilkumar P, Gayatri S, Vinith P, Vignesh V. Extraction of silica from burnt paddy husk. International Journal of ChemTech Research. 2014;**6**(9):4455-4459

[61] Rungrodnimitchai S, Phokhanusai W, Sungkhaho N. Preparation of silica gel from rice husk ash using microwave heating. Journal of Metals, Materials and Minerals. 2017;**19**(2):45-50

[62] Zhang Z, He W, Zheng J, Wang G, Ji J. Rice husk ash-derived silica nanofluids: Synthesis and stability study. Nanoscale Research Letters. 2016;**11**(1):502

[63] Adam F, Chew T-S, Andas J. A simple template-free sol–gel synthesis of spherical nanosilica from agricultural biomass. Journal of Sol-Gel Science and Technology. 2011;**59**(3):580-583

[64] Faizul C, Abdullah C, Fazlul B. Extraction of silica from palm ash via citric acid leaching treatment. Advances in Environmental Biology. 2013a;**7**(12):3690-3695

[65] Carmona V, Oliveira R, Silva W, Mattoso L, Marconcini J. Nanosilica from rice husk: Extraction and characterization. Industrial Crops and Products. 2013;**43**:291-296

[66] Mahmud A, Megat-Yusoff P, Ahmad F, Farezzuan AA. Acid leaching

as efficient chemical treatment for rice husk in production of amorphous silica nanoparticles. ARPN Journal of Engineering and Applied Sciences. 2016;**11**(22):13384

[67] Rafiee E, Shahebrahimi S. Nano silica with high surface area from rice husk as a support for 12-tungstophosphoric acid: An efficient nano catalyst in some organic reactions. Chinese Journal of Catalysis. 2012;**33**(7-8):1326-1333

[68] Bakar RA, Yahya R, Gan SN. Production of high purity amorphous silica from rice husk. Procedia Chemistry. 2016;**19**:189-195

[69] Faizul C, Abdullah C, Fazlul B. Extraction of silica from palm ashvia citric acid leaching treatment. Advances in Environmental Biology. 2013b;**7**(12):3690-3695

[70] Hu S, Hsieh Y-L. Preparation of activated carbon and silica particles from rice straw. ACS Sustainable Chemistry & Engineering. 2014;**2**(4):726-734

[71] Kumar S, Sangwan P, Dhankhar RMV, Bidra S. Utilization of rice husk and their ash: A review. Research Journal of Chemistry and Environment. 2013;**1**(5):126-129

[72] Mor S, Manchanda CK, Kansal SK, Ravindra K. Nanosilica extraction from processed agricultural residue using green technology. Journal of Cleaner Production. 2017;**143**:1284-1290

[73] Akoh H, Tsukasaki Y, Yatsuya S, Tasaki A. Magnetic properties of ferromagnetic ultrafine particles prepared by vacuum evaporation on running oil substrate. Journal of Crystal Growth. 1978;**45**:495-500

[74] Tran PX, Soong Y. Preparation of nanofluids using laser ablation in liquid technique. Pittsburgh, PA, and Morgantown, WV: National Energy Technology Laboratory (NETL); 2007

[75] Zhu H-T, Lin Y-S, Yin Y-S. A novel one-step chemical method for preparation of copper nanofluids. Journal of Colloid and Interface Science. 2004;**277**(1):100-103

[76] Jena P, Brocchi E, Motta M. In-situ formation of Cu–Al_2O_3 nano-scale composites by chemical routes and studies on their microstructures. Materials Science and Engineering: A. 2001;**313**(1-2):180-186

[77] Lee J-H, Hwang KS, Jang SP, Lee BH, Kim JH, Choi SU, et al. Effective viscosities and thermal conductivities of aqueous nanofluids containing low volume concentrations of Al_2O_3 nanoparticles. International Journal of Heat and Mass Transfer. 2008;**51**(11-12):2651-2656

[78] Wei X, Wang L. Synthesis and thermal conductivity of microfluidic copper nanofluids. Particuology. 2010;**8**(3):262-271

[79] Zhu HT, Zhang CY, Tang YM, Wang JX. Novel synthesis and thermal conductivity of CuO nanofluid. The Journal of Physical Chemistry C. 2007;**111**(4):1646-1650

[80] Chen Y, Wang X. Novel phase-transfer preparation of monodisperse silver and gold nanoparticles at room temperature. Materials Letters. 2008;**62**(15):2215-2218

[81] Feng X, Ma H, Huang S, Pan W, Zhang X, Tian F, et al. Aqueous–organic phase-transfer of highly stable gold, silver, and platinum nanoparticles and new route for fabrication of gold Nanofilms at the oil/water Interface and on solid supports. The Journal of Physical Chemistry B. 2006;**110**(25):12311-12317

[82] Yu W, Xie H, Chen L, Li Y. Enhancement of thermal conductivity of kerosene-based Fe_3O_4 nanofluids prepared via phase-transfer method. Colloids and Surfaces A: Physicochemical and Engineering Aspects. 2010;**355**(1-3):109-113

[83] Wang L, Fan J. Nanofluids research: Key issues. Nanoscale Research Letters. 2010;**5**(8):1241

[84] Haddad Z, Abid C, Oztop HF, Mataoui A. A review on how the researchers prepare their nanofluids. International Journal of Thermal Sciences. 2014;**76**:168-189

[85] Fazeli SA, Hashemi SMH, Zirakzadeh H, Ashjaee M. Experimental and numerical investigation of heat transfer in a miniature heat sink utilizing silica nanofluid. Superlattices and Microstructures. 2012;**51**(2):247-264

[86] Pang C, Jung J-Y, Lee JW, Kang YT. Thermal conductivity measurement of methanol-based nanofluids with Al_2O_3 and SiO_2 nanoparticles. International Journal of Heat and Mass Transfer. 2012;**55**(21-22):5597-5602

[87] Bolukbasi A, Ciloglu D. Pool boiling heat transfer characteristics of vertical cylinder quenched by SiO_2–water nanofluids. International Journal of Thermal Sciences. 2011;**50**(6):1013-1021

[88] Darzi AR, Farhadi M, Sedighi K, Shafaghat R, Zabihi K. Experimental investigation of turbulent heat transfer and flow characteristics of SiO_2/water nanofluid within helically corrugated tubes. International Communications in Heat and Mass Transfer. 2012;**39**(9):1425-1434

[89] Yang X-F, Liu Z-H. Pool boiling heat transfer of functionalized nanofluid under sub-atmospheric pressures. International Journal of Thermal Sciences. 2011;**50**(12):2402-2412

[90] Anoop K, Sadr R, Yu J, Kang S, Jeon S, Banerjee D. Experimental study of forced convective heat transfer of nanofluids in a microchannel. International Communications in Heat and Mass Transfer. 2012;**39**(9):1325-1330

[91] Qu J, Wu H. Thermal performance comparison of oscillating heat pipes with SiO_2/water and Al_2O_3/water nanofluids. International Journal of Thermal Sciences. 2011;**50**(10):1954-1962

[92] Hwang Y, Ahn Y, Shin H, Lee C, Kim G, Park H, et al. Investigation on characteristics of thermal conductivity enhancement of nanofluids. Current Applied Physics. 2006;**6**(6):1068-1071

[93] Demirbas MF. Thermal energy storage and phase change materials: An overview. Energy Sources, Part B: Economics, Planning, and Policy. 2006;**1**(1):85-95

[94] Liu Y-D, Zhou Y-G, Tong M-W, Zhou X-S. Experimental study of thermal conductivity and phase change performance of nanofluids PCMs. Microfluidics and Nanofluidics. 2009;**7**(4):579

[95] Shin D, Banerjee D. Enhancement of specific heat capacity of high-temperature silica-nanofluids synthesized in alkali chloride salt eutectics for solar thermal-energy storage applications. International Journal of Heat and Mass Transfer. 2011;**54**(5-6):1064-1070

[96] Wu S, Wang H, Xiao S, Zhu D. Numerical simulation on thermal energy storage behavior of Cu/paraffin nanofluids PCMs. Procedia Engineering. 2012;**31**:240-244

[97] Sharma A. A comprehensive study of solar power in India and world. Renewable and Sustainable Energy Reviews. 2011;**15**(4):1767-1776

[98] Thirugnanasambandam M, Iniyan S, Goic R. A review of solar thermal technologies. Renewable and Sustainable Energy Reviews. 2010;**14**(1):312-322

[99] Kalogirou SA. Solar Energy Engineering: Processes and Systems. United States of America: Academic Press; 2013. ISBN: 9780123972705

[100] Taylor RA, Phelan PE, Otanicar TP, Walker CA, Nguyen M, Trimble S, et al. Applicability of nanofluids in high flux solar collectors. Journal of Renewable and Sustainable Energy. 2011;**3**(2):023104

[101] Badran OO, Abu-Khader MM. Evaluating thermal performance of a single slope solar still. Heat and Mass Transfer. 2007;**43**(10):985-995

[102] Kianifar A, Heris SZ, Mahian O. Exergy and economic analysis of a pyramid-shaped solar water purification system: Active and passive cases. Energy. 2012;**38**(1):31-36

[103] Gnanadason MK, Kumar PS, Rajakumar S, Yousuf MS. Effect of nanofluids in a vacuum single basin solar still. International Journal of Advanced Engineering Research and Studies. 2011;**1**:171-177

[104] Nijmeh S, Odeh S, Akash B. Experimental and theoretical study of a single-basin solar sill in Jordan. International Communications in Heat and Mass Transfer. 2005;**32**(3-4):565-572

[105] Shanthi R, Anandan SS, Ramalingam V. Heat transfer enhancement using nanofluids. Thermal Science. 2012;**16**(2):423-444

[106] Tzeng S-C, Lin C-W, Huang K. Heat transfer enhancement of nanofluids in rotary blade coupling of four-wheel-drive vehicles. Acta Mechanica. 2005;**179**(1-2):11-23

[107] Popa I, Gillies G, Papastavrou G, Borkovec M. Attractive and repulsive electrostatic forces between positively charged latex particles in the presence of anionic linear polyelectrolytes. The Journal of Physical Chemistry B. 2010;**114**(9):3170-3177

[108] Kao M-J, Chang H, Wu Y-Y, Tsung T-T, Lin H-M. Producing aluminum-oxide brake nanofluids using plasma charging system. Journal of the Chinese Society of Mechanical Engineers. 2007a;**28**(2):123-131

[109] Kao M, Lo C, Tsung T, Wu Y, Jwo C, Lin H. Copper-oxide brake nanofluid manufactured using arc-submerged nanoparticle synthesis system. Journal of Alloys and Compounds. 2007;**434**:672-674

[110] Yu W, Xie H. A review on nanofluids: Preparation, stability mechanisms, and applications. Journal of Nanomaterials. 2012;**2012**:435873

[111] Tsai C, Chien H, Ding P, Chan B, Luh T, Chen P. Effect of structural character of gold nanoparticles in nanofluid on heat pipe thermal performance. Materials Letters. 2004;**58**(9):1461-1465

[112] Ma H, Wilson C, Borgmeyer B, Park K, Yu Q, Choi S, et al. Effect of nanofluid on the heat transport capability in an oscillating heat pipe. Applied Physics Letters. 2006;**88**(14):143116

[113] Lin Y-H, Kang S-W, Chen H-L. Effect of silver nano-fluid on pulsating heat pipe thermal performance. Applied Thermal Engineering. 2008;**28**(11-12):1312-1317

[114] Vafaei S, Borca-Tasciuc T, Podowski M, Purkayastha A, Ramanath G, Ajayan P. Effect of nanoparticles on sessile droplet contact angle. Nanotechnology. 2006;**17**(10):2523

[115] Zhang L, Jiang Y, Ding Y, Povey M, York D. Investigation into the antibacterial behaviour of suspensions of ZnO nanoparticles (ZnO nanofluids). Journal of Nanoparticle Research. 2007;**9**(3):479-489

[116] Jalal R, Goharshadi EK, Abareshi M, Moosavi M, Yousefi A, Nancarrow P. ZnO nanofluids: Green synthesis, characterization, and antibacterial activity. Materials Chemistry and Physics. 2010;**121**(1-2):198-201

[117] Panáček A, Kvitek L, Prucek R, Kolář M, Večeřová R, Pizúrová N, et al. Silver colloid nanoparticles: Synthesis, characterization, and their antibacterial activity. The Journal of Physical Chemistry B. 2006;**110**(33):16248-16253

[118] Lyon DY, Alvarez PJ. Fullerene water suspension (nC60) exerts antibacterial effects via ROS-independent protein oxidation. Environmental Science & Technology. 2008;**42**(21):8127-8132

[119] Micali F, Milanese M, Colangelo G, de Risi A. Experimental investigation on 4-strokes biodiesel engine cooling system based on nanofluid. Renewable Energy. 2018;**125**:319-326

[120] Al-Jethelah M, Tasnim SH, Mahmud S, Dutta A. Nano-PCM filled energy storage system for solar-thermal applications. Renewable Energy. 2018;**126**:137-155

[121] Said Z, Assad MEH, Hachicha AA, Bellos E, Abdelkareem MA, Alazaizeh DZ, et al. Enhancing the performance of automotive radiators using nanofluids. Renewable and Sustainable Energy Reviews. 2019;**112**:183-194

[122] Akilu S, Baheta AT, Said MAM, Minea AA, Sharma K. Properties of glycerol and ethylene glycol mixture based SiO_2-CuO/C hybrid nanofluid for enhanced solar energy transport. Solar Energy Materials and Solar Cells. 2018;**179**:118-128

[123] Yao S, Huang X, Song Y, Shen Y, Zhang S. Effects of nanoparticle types and size on boiling feat transfer performance under different pressures. AIP Advances. 2018;**8**(2):025005

Chapter 5

Preparation of Nano-Particles and Their Applications in Adsorption

Tooba Saeed, Abdul Naeem, Tahira Mahmood and Nazish Huma Khan

Abstract

The nano-technologies and nano-materials draw incredible consideration in recent years. Nano-particles are the particles having size ranging from 1 to 100 nm. The nano-particles are usually categorized into different classes, and their classification is based on size, shape, material production, and dimension. They show superior properties, i.e., enhanced reactivity, high BET surface area, sensitiveness, and steadiness as compared to their bulk materials. In this chapter, different approaches of synthesizing nano-particles, including sol gel, chemical vapor deposition, and biosynthesis are talked over. In the treatment of wastewater, nano-particles offer a possibility for effective adsorption of contaminants organic as well as inorganic. This chapter presents an overview on nano-particles, their types, characteristics, synthetic approaches, and applications in the field of surface chemistry.

Keywords: nano-particles, adsorption, sol–gel, chemical vapor deposition, biosynthesis, carbon nano-tubes, mechanical milling, nano-lithography, laser ablation, iron nano-particles, manganese oxide nano-particles, zinc oxide and magnesium oxide

1. Introduction

This chapter consists of three main sections. The first section gives an overview about the introduction of nano-particles. The next section is about the synthesis of nanoparticles and the last section describes the use of nanoparticles as adsorbents.

The preface "nano" is known for nineteenth century for its ever-increasing applications in various fields of science. A few nano-containing terms that are found in the record (usually in scientific reports and books) are nano-materials, nano-chemistry, nano-science or nanotechnology. The preface nano comes from a Latin nanos meaning dwarf that means extremely small. According to units system working internationally, it is used to represent a reduction factor of 109 times. Consequently, the nano-materials are usually dignified in nano-meters (1 nm is equivalent to 10^{-9} m) and it comprises systems having size less than macroscopic measurements and greater than molecular ones (mostly >1 nm and <100 nm) at least in one spatial dimension. This characteristic scale might be used for a particle size, diameter and layer thickness [1–5].

1.1 Classification of nano-materials

The nano-materials are different in structure, size and shape. They can be of various shapes like rod, globular, conical, hollow, coiled, plane, cylindrical and asymmetrical, while some are crystalline or amorphous.

Nano-materials are generally classified into nano-emulsions, nano-clays and nano-particles. Nano-particles are present as nano-composites or nano-structures. These nano-structures are made from basic units or blocks having small dimensionality i.e. zero, one, two and three dimensions. In zero dimensional nano-particles, the moment of electrons is cramped in all three dimensions, e.g. quantum dots. If electrons can move freely in x-direction only, they are one dimensional nanoparticles e.g. quantum wires. Whereas, in two dimensional thin films and three dimensional nano-structured materials, free electrons can move freely in x, y and x, y, z directions respectively.

Based on material production and role in sorption process, nano-particles can also be categorized into organic, mixed oxide nano-structures, magnetic, inorganic (metallic) and carbon based nano-particles. Organic nano-particles are self-assembled, three dimensional fabricated by synthetic and natural organic molecules, i.e. protein masses, milk suspension and lipid bulks etc. Commonly known organic nanoparticles are micelles, dendrimers, ferritin and liposomes. Inorganic nanoparticles usually are manufactured from inorganic salt precipitations. They are non-carbon containing particles and their most common examples are metal and metal oxide particles. Carbon based nano-particles are manufactured entirely from carbon e.g. graphene, fullerenes, carbon nano-tubes, carbon nano-fibers, carbon black and sometimes activated carbon (**Figure 1**) [6–10].

In nano-materials (especially nano-particles) molecules and atoms act differently and reveal inimitable physical, chemical and electronic properties. These properties are different from their bulk counterparts and sometimes the same kind of nano-particles can show diverse characters. Physical properties of nano-particles

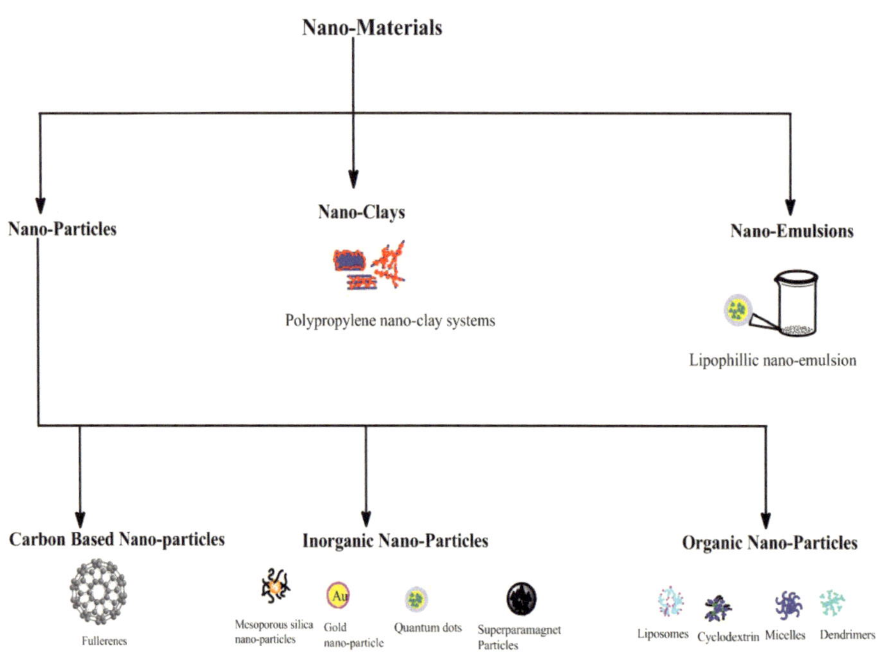

Figure 1.
Classification of nano-materials.

include absorption, reflection, light dispersion, color of nano-particles, hydrophobicity, hydro-philicity, suspension and dispersion. When layered onto a surface or in the form of solution their absorption and reflection properties make them a perfect choice for different fields. They also show outstanding chemical properties, i.e. anti-destructive, oxidation, reduction, flammability, sensitivity and stability towards humidity, atmosphere, heat, light and dis-infection, non-toxicity, biodegradability, anti-bacterial and fungal properties. These properties also enable them ideal materials for environmental and biomedical applications. Nano-particles also exhibit mechanical properties like elasticity, ductility, flexibility, tensile strength and electrical properties including semi-conductivity, conductivity and resistivity which have directed a route for them to be used in renewable energy applications.

These distinctive and inimitable properties of nano-particles make them perfect and formidable for amazing and interesting applications in physical science, material science, agriculture, food, engineering, industrial and biomedical sciences (**Figure 2**).

These applications encompass them to be used in electronic, drug delivery, optical, mechanics, catalysis, bio-encapsulation and wastewater treatment especially adsorption [4–15]. Besides all these properties, nano-particles have some toxic effects for aquatic and human health. As they have small size, they can easily enter through the skin of organisms and consequently enter into the body fluid. Furthermore, nano-particles used in sun screens can absorb deep inside and become toxic to the skin, bones and liver cells. Their greater surface area often makes them more sensitive, explosive and reactive. Inhaling directly in the environment of nano-particles can adversely affect the function of lungs especially in human being. However, highlighting the health issues caused by nano-particles, does not mean to ignore their extraordinary importance in technology, industries and environment [16–18].

It is important to mention that water is one among the basic necessities of every organism. From total water present on earth, only 0.01% portion is available for human [19, 20]. Shortage of drinking water is increasing day by day due to demolition of water means [21].

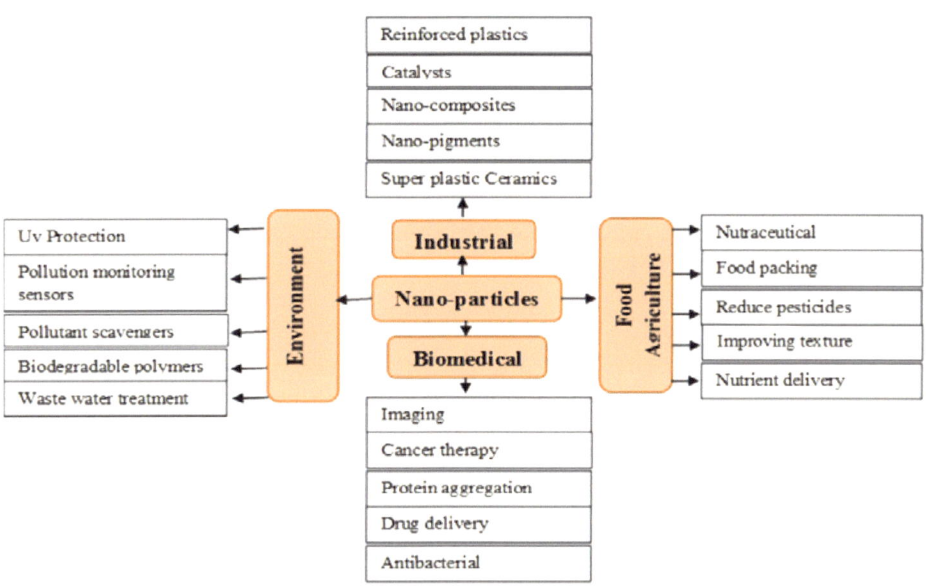

Figure 2.
Applications of nano-particles in different fields [20].

The main sources of water contamination are agricultural, industrial and domestic effluents. Industries can help on the one hand in the development of economy, whereas, on the other hand, they are mainly responsible for various environmental issues i.e. water, air and soil pollution. Drinking water containing agricultural and industrial effluents cause different diseases such as, cancer, eye irritation, dermatitis, cell damage and dysfunction of kidney, respiratory and reproductive system even in a very insignificant quantity. Hence, treatment methods for drinking as well as wastewater are one of the most important requirements for emerging and growing health and economy. Various techniques have been used for decontamination of pollutants from industrial wastewater, including reverse-osmosis, ion-exchange, chemical oxidation, flocculation or coagulation and precipitation. Each technique has its individual disadvantages as they are energy dependent, economically as well as technically not sound and achievable. Literature exhibited that from all these treatment techniques, adsorption is one of the most effective technique for water decontamination. Adsorption is simple, adaptable, highly potential, efficient and recyclable technique [22, 23]. A range of effective, low-cost and environment friendly nano-materials with outstanding properties have been developed for prospective applications in decontamination of industrial effluents, surface, ground and drinking water. Literature also revealed that nanoparticles behave as an ideal adsorbent as they are environmentally benign, selective, efficient, recyclable, high surface area and maximum adsorption capacity even at a very low concentration [24–26].

The recent progress related to the different aspects of adsorption using nanoparticles have described in several reviews and book chapters. This chapter focuses on the various techniques used for preparation of nano-particles and their applications in the field of adsorption.

2. Preparation of nano-particles

The nano-particles can be prepared by various processes divided into i.e. bottom up and top down techniques. Bottom up methods include the reduction of material components up to the atomic level and then with further self-assembly lead to the formation of nano-particles. However, during self-assembly, the physical forces functioning at nano-scale are used to connect basic units into macro structures. Pyrolysis, bio-synthesis, sole gel, spinning and chemical vapor deposition are most extensively used methods fall in this approach [6, 27]. Whereas, top down techniques including sputtering, laser ablation, nano-lithography, mechanical milling and thermal decomposition, starting with a pattern produced on a higher scale, then compacted to nanoscale. Both of these techniques are contradictory and schematically represented in **Figure 3**.

2.1 Bottom up technique

2.1.1 Sol gel

The sol is a colloid where the aggregates of fine particle are distributed in liquid phase. They are larger in size ranging from 1 nm to 1 μm than nano-particles. Whereas, solid macromolecules immersed in a solvent, called as gel. Sol gel is one of the simplest and, most commonly used method for the synthesis of nano-particles. It is a chemical method which comprises of a solution working as a precursor for an assimilated system of distinct particles. In this method, metal oxides, metal chlorides and alkoxysilanes (typically tetramethoxy and ethoxysilanes) are most

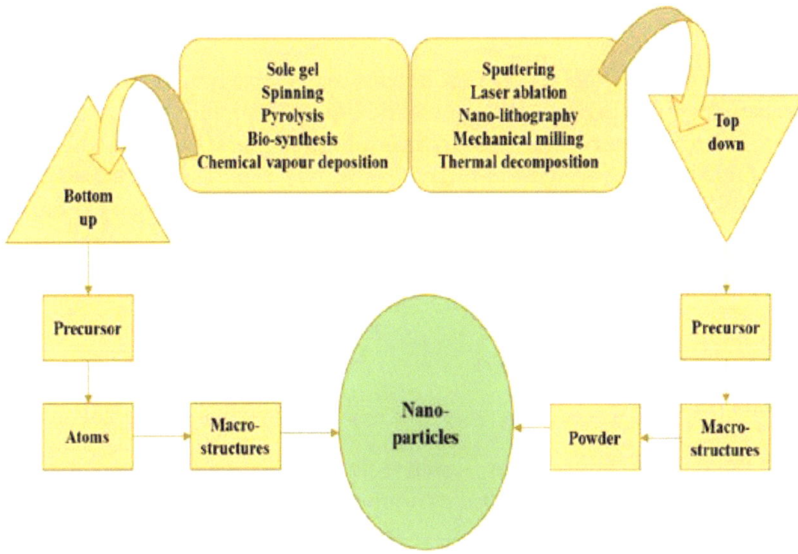

Figure 3.
Schematic representation of bottom up and top down techniques.

commonly used as precursors. The precursor is mixed by means of mixing, quivering sonicating or stirring and is then spread in second liquid which form a solid–liquid phase. Catalyst is commonly used to start the reaction and to control the pH of the system. Sedimentation, filtration and centrifugation are the typical methods used for phase separation to get nano-particles and then the sample is dried to remove moistness. The main advantages of this process are to attain uniform nano-structures even at a very low temperature, having controlled chemical composition and purity [6, 27–30]. This process is not easily scalable having different drying steps involved as well as it is difficult to control synthesis during this process.

2.1.2 Chemical-vapor deposition (CVD)

In this method of preparation, substrate is coated with a thin film of gaseous reactants. The gas molecules are combined at ambient temperature in a reaction chamber to carry out deposition. Upon heating substrate comes in vicinity of combined gas where a chemical reaction occurs and a thin film is formed on the surface of substrate. This thin film can be recovered and reused for different applications. The basic influencing factor in this method is the temperature of the substrate. The nanoparticles achieved through this method are highly pure, uniform in size, strong and have high mechanical stability. The disadvantages of CVD include the use of special equipment as well as the high toxicity of the gaseous by-products [8, 31, 32].

2.1.3 Biosynthesis

Biosynthesis is one of the inexpensive, green, safe, decomposable and environment friendly methods used for the synthesis of nano-particles. In this method bacterium, fungi and plant extracts are used in conjunction with precursor for bio-reduction and capping functions rather than conventional chemicals. This method has its distinctive and enriched properties that find its approaches in medical applications [33, 34].

2.1.4 Pyrolysis

Pyrolysis is the method used in industries to prepare nan-particles on large scale. In this method, precursor used can be liquid or vapor. A furnace is used in order to burn the precursor. Precursor is added to the furnace through a small opening where flame is applied to burn it. [13]. Nano-particles are collected by the gases produced as by product. Pyrolysis is an effective method for nano-particle preparation due to its simplicity, high product yield and sensitiveness [35, 36].

2.2 Top down techniques

2.2.1 Mechanical milling

One of the most widely used top down techniques to produce nano-particles is mechanical milling. In this method various elements are milled under an inert atmosphere and during this process particles are milled and post annealed. The influencing factor in this method is plastic distortion which end up with particle size, breakage that ends up in particle size, and cold-soldering that ends up to increased particle size [37–39].

2.2.2 Nano-lithography

Nanolithography is the investigation of manufacturing nano-scale structures of one dimension at least, with size ranging from 1 to 100 nm. There are different nano-lithographic forms, for example optical, electron-pillar, multiphoton, nanoimprint and filtering test lithography. Mostly lithography is the way towards printing a required shape or structure of a light sensitive material, which specifically evacuates a bit of material to make the ideal shape and structure. The primary advantage of nanolithography is to create a bunch from a solitary nano-particle with desired shape and size [40–42]. A sophisticated equipment is required in this method which is cost effective.

2.2.3 Laser-ablation

Laser-ablation is a typical method for the preparation of nano-particles from various solvents in solution. A metal immersed in a liquid solution is irradiated by the laser beam, resulting in the formation of plasma crest that yields into nano-particles. In this process, a chemical reduction of metals occurs to produce inorganic (metal based) nano-particles. As laser ablation gives a steady synthesis of nano-particles in natural solvents and water that does not require any balancing agent or synthetic substance. It is a 'green' process and its setup is shown in **Figure 4** [43–45].

2.2.4 Thermal decomposition

In this method heat is applied to decompose the chemical bonds of the compound. It is an endothermic chemical process where the nano-particles are synthesized by rotting a metal at a precise temperature called as decomposition temperature. As a result of this decomposition secondary products are also produced. This method is useful for the preparation of metal oxide and carbon based nano-particles [46, 47].

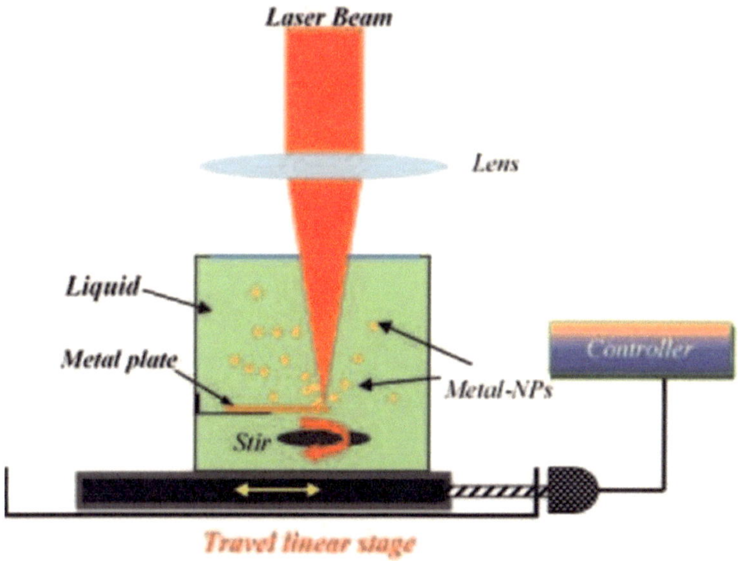

Figure 4.
Laser ablation setup [43].

3. Applications of nano-particles as adsorbent

Now a days, one of the foremost problems that is facing by the world is accessibility of clean drinking water. Demand for fresh and clean water is increasing day by day due to increasing population. In developing and industrialized countries, clean water deficiency is intensified by human as well as by the industrial effluents (metals and dyes). These effluents are directly discharged into water bodies and contaminate them. As described in introduction part of this chapter, sorption is declared to be one of the best and suitable methods for wastewater treatment [48–50].

The sorption method is a surface phenomenon during which sorbate is gathered on the sorbent surface. When adsorbate molecules from solution come to the vicinity of adsorbent surface, then some of the molecules adsorb onto the sorbent surface by intermolecular forces of attraction between surfaces of adsorbent and adsorbate molecules. The particular nature of interaction can be determined by the type of species concerned. However, the sorption method is usually classified as physi-sorption where the sorbate bound on the sorbent surface through valence or electrostatic bonding and chemi-sorption where molecule attached through chemical bonding [51–54].

Nano-particles have a high specific surface area, sorption active sites, solubility, efficiency and fractal dimension, short intra-particle diffusion distance, well defined chemical composition, small particle size and tunable pore size as compare to the their bulk counterparts that are responsible for their valuable features for effective sorption especially chemical activity and fine grain size. The high surface area and high sorption active site in nano-particles are due to high surface-energy and size dependent surface structure at nano-scale. The nano-particles have the highest efficiency towards sorption of organic and inorganic pollutants and their selectivity towards contaminants can be increased by functionalizing the surface of nano-particles. Iron oxide, titanium dioxide, manganese dioxide, silica nanoparticles, alumina, zinc oxide, dicalcium phosphate, copper, silver, maghemite, gold nano-particles, etc. are discovered as cheap, efficient, easy to synthesize and environment friendly sorbents for the removal of pollutants. Among nano-particles (metal oxides), the magnetic nano-particles have acquired a substantial importance due to their interesting magnetic properties i.e. super para-magnetism, strong reaction even at minor applied magnetic field [6, 12, 55, 56].

Moreover, a recent improvement on carboniferous and siliceous nano-materials enclosed nano-sheets, nano-tubes and nano-particles of carbon and silicon declared as efficient adsorbents for sorption of metals and dyes from wastewater. Some oxides and carbon based nanomaterials are discussed below [55–57].

3.1 Iron nano-particles

Iron based nano-particles are most commonly used adsorbents for the removal of toxic materials from aqueous solutions. These nano-particles are declared as most efficient, cost effective and ecofriendly sorbent with less chance for the production of secondary contaminants. The adsorption process by iron oxide nano-particles is affected by pH, temperature, adsorbent dosage and equilibrium time. Modification of these materials increased their surface properties for the removal of metals, i.e. cadmium(II), lead(II), copper(II), chromium(II), nickel(II), arsenic(III) and anionic and cationic dyes [58–60].

3.2 Manganese oxide nano-particles

Manganese oxide nano-particles have a high specific surface area which makes them effective adsorbent for the removal of heavy metals i.e. arsenic(III), lead(II), cadmium(II) and ionic dyes. Manganese oxide nano-particles can also be modified into hydrous manganese oxide, nano-porous and nano-tunnel manganese oxide to improve their surface area and porosity for excellent adsorption [61, 62].

3.3 Zinc oxide

These are porous micro nano-structure with high Brunauer-Emmett-Teller (BET) surface area. Most widely used nano-sorbents of zinc oxide are nano-assembled, nano-sheets, nano-rods, nano-plates and micro-spheres for competent removal of dyes and inorganic pollutants from aqueous phase. Whereas, micro-porous nano assemblies of zinc oxide display maximum potential for the removal of lead(II), arsenic(III) and mercury(II) because of their electro-positive nature [63–65].

3.4 Magnesium oxide

The sorption capacity of the magnesium oxide nano-particles is much greater than its bulk counterpart. Their micro-spheres are innovative structure, with increased sorption capacity for the sorption of heavy metals. Various modification of magnesium oxide nano-particles i.e. nano-rods, nano-tubes, nano-wires has been reported as improved sorption affinity towards metals and organic effluents [66–68].

3.5 Carbon nano-tubes

They are the most widely used material for the sorption of heavy metals as well as organic dyes from aqueous media. Though, they have meager dispersal capacity, very small size of particles and separation complications are some difficulties for using carbon nano-tubes as sorbents. Whereas, these difficulties can be overawed by modifying carbon nano-tubes into multi walled carbon nano-tubes. Literature revealed that the multi walled nano tubes and alumina supported carbon nano-tubes, more competently removed metals such as Mn(II), Cu(II) and Pb(II) more efficiently as compare to unmodified material [22, 69, 70].

4. Conclusion

Nano-technology is refining our everyday life by increasing the proficiency and purity of many substances. As described in this chapter, there are different techniques for the synthesis of nano-particles, but laser ablation chemical vapor deposition, nano-lithography, biosynthesis, mechanical milling, and sol–gel are the most suitable techniques because they are less time consuming methods. Nano-particles with inimitable chemical and physical characteristics, have a remarkable prospective for the adsorption of contaminant but still their applications for wastewater treatment are inadequate. However, nanoparticles have pronounced future due to their proficiency and environmentally benign property.

Acknowledgements

The authors would like to convey their gratefulness to National Centre of Excellence in Physical Chemistry, University of Peshawar for providing us necessary support and facilities to carry out this study.

Author details

Tooba Saeed[1*], Abdul Naeem[1], Tahira Mahmood[1] and Nazish Huma Khan[2]

1 National Centre of Excellence in Physical Chemistry, University of Peshawar, Peshawar, Pakistan

2 Department of Environmental Sciences, University of Sawabi, Sawabi, Pakistan

*Address all correspondence to: khantooba590@gmail.com

References

[1] Hashim PW, Nia JK, Han G, Ratner D. Nanoparticles in dermatologic surgery. Journal of the American Academy of Dermatology. 2019;**1**:1-19. DOI: 10.1016/j.jaad.2019.04.020

[2] Rashidi S, Karimi N, Mahian O, Abolfazli Esfahani J. A concise review on the role of nanoparticles upon the productivity of solar desalination systems. Journal of Thermal Analysis and Calorimetry. 2019;**135**:1145-1159. DOI: 10.1007/s10973-018-7500-8

[3] Namita R. Methods of preparation of nanoparticles—A review. African Journal of Biotechnology. 2013;**12**:32-41. DOI: 10.1155/2010/745120

[4] Luo XL, Morrin A, Killard AJ, Smyth MR. Application of nanoparticles in electrochemical sensors and biosensors. Electroanalysis. 2006;**18**:319-326. DOI: 10.1002/elan.200503415

[5] Graham B, O'Malley W, Stephan H, Barreto JA, Spiccia L, Kubeil M. Nanomaterials: Applications in cancer imaging and therapy. Advanced Materials. 2011;**23**:H18-H40. DOI: 10.1002/adma.201100140

[6] Konwar R, Ahmed AB. Nanoparticle: An overview of preparation, characterization and application. International Research Journal of Pharmacy. 2016;**4**:47-57. DOI: 10.7897/2230-8407.04408

[7] Mageswari A, Srinivasan R, Subramanian P, Ramesh N, Gothandam KM. Nanoscience in Food and Agriculture. Vol. 26. Springer; 2016. pp. 31-71. DOI: 10.1007/978-3-319-48009-1_2

[8] Sahu RK, Hiremath SS. A review on the classification, characterisation, synthesis of nanoparticles and their application related content synthesis of aluminium nanoparticles in a water/polyethylene glycol mixed solvent using-EDM. IOF Conference Series Materials Science and Engineering. 2017;**263**:19-32. DOI: 10.1088/1757-899X/263/3/032019

[9] Durairaj S. Application of nanoparticles in wastewater treatment. Pollution Research. 2017;**33**:567-571. DOI: 10.1016/j.jece.2017.05.029

[10] Blanco-López MC, Rivas M. Nanoparticles for bioanalysis. Analytical and Bioanalytical Chemistry. 2019;**411**:1789-1790. DOI: 10.1007/s00216-019-01680-x

[11] Martin CR. Nanomaterials: A membrane-based synthetic approach. Science (80). 1994;**266**:1961-1966. DOI: 10.1126/science.266.5193.1961

[12] Horri BA, Abdullah AZ, Tan KB, VakiliM,SalamatiniaB,PohPE.Adsorption of dyes by nanomaterials: Recent developments and adsorption mechanisms. Separation and Purification Technology. 2015;**150**:229-242. DOI: 10.1016/j.seppur.2015.07.009

[13] Tenne R. Fullerene-like materials and nanotubes from inorganic compounds with a layered (2-D) structure. Colloids and Surfaces A: Physicochemical and Engineering Aspects. 2002;**208**:83-92. DOI: 10.1016/S0927-7757(02)00104-8

[14] Khan R, Qadeer A, Ahmad E, Zaman M, Rabbani G. Nanoparticles in relation to peptide and protein aggregation. International Journal of Nanomedicine. 2014;**9**:899-912

[15] Fawole OG, Cai XM, Mackenzie AR.Gas flaring and resultant air pollution: A review focusing on black carbon. Environmental Pollution. 2016;**216**:182-197. DOI: 10.1016/j.envpol.2016.05.075

[16] Gwinn MR, Vallyathan V. Nanoparticles: Health effects—Pros and cons. Environmental Health Perspectives. 2006;**114**:1818-1825. DOI: 10.1289/ehp.8871

[17] Brook RD, Franklin B, Cascio W, Hong Y, Howard G, Lipsett M, et al. Air pollution and cardiovascular disease: A statement for healthcare professionals from the expert panel on population and prevention science of the American Heart Association. Circulation. 2004;**109**:2655-2671. DOI: 10.1161/01.CIR.0000128587.30041.C8

[18] Bharali DJ, Klejbor I, Stachowiak EK, Dutta P, Roy I, Kaur N, et al. Organically modified silica nanoparticles: A nonviral vector for in vivo gene delivery and expression in the brain. Nature Methods. 2005;**2**:639. DOI: 10.1038/nmeth0905-639

[19] Ryu CH, Joo SJ, Kim HS. Two-step flash light sintering of copper nanoparticle ink to remove substrate warping. Applied Surface Science. 2016;**384**:182-191. DOI: 10.1016/j.apsusc.2016.05.025

[20] Azizullah A, Khattak MNK, Richter P, Häder DP. Water pollution in Pakistan and its impact on public health—A review. Environment International. 2011;**37**:479-497

[21] Carolin CF, Kumar PS, Saravanan A, Joshiba GJ, Naushad M. Efficient techniques for the removal of toxic heavy metals from aquatic environment: A review. Journal of Environmental Chemical Engineering. 2017;**5**:2782-2799. DOI: 10.1016/j.jece.2017.05.029

[22] Saleh TA, Gupta VK. Column with CNT/magnesium oxide composite for lead (II) removal from water. Environmental Science and Pollution Research. 2012;**19**:1224-1228. DOI: 10.1007/s11356-011-0670-6

[23] Hua M, Zhang S, Pan B, Zhang W, Lv L, Zhang Q. Heavy metal removal from water/wastewater by nanosized metal oxides: A review. Journal of Hazardous Materials. 2012;**211-212**:317-331. DOI: 10.1016/j.jhazmat.2011.10.016

[24] Sadegh H, Ali GAM, Gupta VK, Makhlouf ASH, Shahryari-ghoshekandi R, Nadagouda MN, et al. The role of nanomaterials as effective adsorbents and their applications in wastewater treatment. Journal of Nanostructure in Chemistry. 2017;**7**:1-14. DOI: 10.1007/s40097-017-0219-4

[25] Ali R, Mahmood T, Ud Din S, Naeem A, Aslam M, Farooq M. Efficient removal of hazardous malachite green dye from aqueous solutions using H_2O_2 modified activated carbon as potential low-cost adsorbent: Kinetic, equilibrium, and thermodynamic studies. Desalination and Water Treatment. 2019;**151**:167-182. DOI: 10.5004/dwt.2019.23813

[26] Nasrullah A, Bhat AH, Isa MH, Danish M, Naeem A, Muhammad N, et al. Efficient removal of methylene blue dye using mangosteen peel waste: Kinetics, isotherms and artificial neural network (ANN) modelling. Desalination and Water Treatment. 2017;**86**:191-202. DOI: 10.5004/dwt.2017.21295

[27] Bodnar M, Hartmann JF, Borbely J. Preparation and characterization of chitosan-based nanoparticles. Biomacromolecules. 2005;**6**:2521-2527. DOI: 10.1021/bm0502258

[28] Ramesh S. Sol-gel synthesis and characterization of nanoparticles. Journal of Nanoscience. 2013;**2013**:1-8. DOI: 10.1155/2013/929321

[29] Mann S, Burkett SL, Davis SA, Fowler CE, Mendelson NH, Sims SD, et al. Sol-gel synthesis of organized matter. Chemistry of Materials.

1997;**9**:2300-2310. DOI: 10.1021/cm970274u

[30] Lu CH, Jagannathan R. Cerium-ion-doped yttrium aluminum garnet nanophosphors prepared through sol-gel pyrolysis for luminescent lighting. Applied Physics Letters. 2002;**80**:3608-3610. DOI: 10.1063/1.1475772

[31] Bhaviripudi S, Mile E, Steiner SA, Zare AT, Dresselhaus MS, Belcher AM, et al. CVD synthesis of single-walled carbon nanotubes from gold nanoparticle catalysts. Journal of the American Chemical Society. 2007;**129**:1516-1517. DOI: 10.1021/ja0673332

[32] Li Y, Liu J, Wang Y, Wang ZL. Preparation of monodispersed Fe-Mo nanoparticles as the catalyst for CVD synthesis of carbon nanotubes. Chemistry of Materials. 2001;**13**:1008-1014. DOI: 10.1021/cm000787s

[33] Kulkarni N, Muddapur U. Biosynthesis of metal nanoparticles: A review. Journal of Nanotechnology. 2014;**2014**:1-8. DOI: 10.1155/2014/510246

[34] Shah M, Fawcett D, Sharma S, Tripathy SK, Poinern GEJ. Green synthesis of metallic nanoparticles via biological entities. Materials. 2015;**8**:7278-7308. DOI: 10.3390/ma8115377

[35] D'Amato R, Falconieri M, Gagliardi S, Popovici E, Serra E, Terranova G, et al. Synthesis of ceramic nanoparticles by laser pyrolysis: From research to applications. Journal of Analytical and Applied Pyrolysis. 2013;**104**:461-469. DOI: 10.1016/j.jaap.2013.05.026

[36] Johannessen T, Jensen JR, Mosleh M, Johansen J, Quaade U, Livbjerg H. Applications in catalysis and product/process engineering. Chemical Engineering Research and Design. 2004;**82**:1444-1452. DOI: 10.1205/cerd.82.11.1444.52025

[37] Prasad Yadav T, Manohar Yadav R, Pratap Singh D. Mechanical milling: A top down approach for the synthesis of nanomaterials and nanocomposites. Nanoscience and Nanotechnology. 2012;**2**:22-48. DOI: 10.5923/j.nn.20120203.01

[38] Catalent. Mechanical milling. Catalent. 1996;**19**:1-2. DOI: 10.1016/j.cherd.2019.06.029

[39] Mucsi G. A review on mechanical activation and mechanical alloying in stirred media mill. Chemical Engineering Research and Design. 2019;**148**:460-474. DOI: 10.1016/j.cherd.2019.06.029

[40] Seisyan RP. Nanolithography in microelectronics: A review. Technical Physics. 2011;**56**:1061-1073. DOI: 10.1134/s1063784211080214

[41] Wang L, Major D, Paga P, Zhang D, Norton MG, McIlroy DN. High yield synthesis and lithography of silica-based nanospring mats. Nanotechnology. 2006;**17**(11). DOI: 10.1088/0957-4484/17/11/S12

[42] Hassani SS, Sobat Z. Studying of various nanolithography methods by using scanning probe microscope. International Journal of Nano Dimension. 2011;**1**:159-175

[43] Sadrolhosseini AR, Mahdi MA, Alizadeh F, Rashid SA. Laser Technology and Its Applications. IntechOpen; 2018. pp. 117-121. DOI: 10.5772/57353

[44] Amendola V, Meneghetti M. Laser ablation synthesis in solution and size manipulation of noble metal nanoparticles. Physical Chemistry Chemical Physics. 2009;**11**:3805-3821. DOI: 10.1039/b900654k

[45] Kabashin AV, Meunier M. Synthesis of colloidal nanoparticles during femtosecond laser ablation of gold in water. Journal of Applied Physics. 2003;**94**:7941. DOI: 10.1063/1.1626793

[46] Salavati-Niasari M, Davar F, Mir N. Synthesis and characterization of metallic copper nanoparticles via thermal decomposition. Polyhedron. 2008;**27**:3514-3518. DOI: 10.1016/j.poly.2008.08.020

[47] Ealias AM, Saravanakumar MP. A review on the classification, characterisation, synthesis of nanoparticles and their application. IOP Conference Series: Materials Science and Engineering. 2017;**263**:1-9. DOI: 10.1088/1757-899X/263/3/032019

[48] Hameed BH, Ahmad AA. Batch adsorption of methylene blue from aqueous solution by garlic peel, an agricultural waste biomass. Journal of Hazardous Materials. 2009;**164**:870-875. DOI: 10.1016/j.jhazmat.2008.08.084

[49] Arami M, Limaee NY, Mahmoodi NM, Tabrizi NS. Removal of dyes from colored textile wastewater by orange peel adsorbent: Equilibrium and kinetic studies. Journal of Colloid and Interface Science. 2005;**288**:371-376. DOI: 10.1016/j.jcis.2005.03.020

[50] Mittal A, Mittal J, Malviya A, Gupta VK. Adsorptive removal of hazardous anionic dye "Congo red" from wastewater using waste materials and recovery by desorption. Journal of Colloid and Interface Science. 2009;**340**:16-26. DOI: 10.1016/j.jcis.2009.08.019

[51] Guesmi Y, Agougui H, Lafi R, Jabli M, Hafiane A. Synthesis of hydroxyapatite-sodium alginate via a co-precipitation technique for efficient adsorption of methylene blue dye. Journal of Molecular Liquids. 2018;**249**:912-920. DOI: 10.1016/j.molliq.2017.11.113

[52] Tamez Uddin M, Rukanuzzaman M, Maksudur Rahman Khan M, Islam MA. Adsorption of methylene blue from aqueous solution by jackfruit (*Artocarpus heteropyllus*) leaf powder: A fixed-bed column study. Journal of Environmental Management. 2009;**90**:3443-3450. DOI: 10.1016/j.jenvman.2009.05.030

[53] Iryani A, Ilmi MM, Hartanto D. Adsorption study of Congo red dye with ZSM-5 directly synthesized from bangka kaolin without organic template. Malaysian Journal of Fundamental and Applied Sciences. 2018;**13**:832-839. DOI: 10.11113/mjfas.v13n4.934

[54] Rath SS, Singh S, Rao DS, Nayak BB, Mishra BK. Adsorption of heavy metals on a complex Al–Si–O bearing mineral system: Insights from theory and experiments. Separation and Purification Technology. 2017;**186**:28-38. DOI: 10.1016/j.seppur.2017.05.052

[55] Ibrahim RK, Hayyan M, AlSaadi MA, Hayyan A, Ibrahim S. Environmental application of nanotechnology: Air, soil, and water. Environmental Science and Pollution Research. 2016;**23**:13754-13788. DOI: 10.1007/s11356-016-6457-z

[56] Ortega PFR, Trigueiro JPC, Santos MR, Denadai ÂML, Oliveira LCA, Teixeira APC, et al. Thermodynamic study of methylene blue adsorption on carbon nanotubes using isothermal titration calorimetry: A simple and rigorous approach. Journal of Chemical & Engineering Data. 2017;**62**:729-737. DOI: 10.1021/acs.jced.6b00804

[57] Lu H, Wang J, Stoller M, Wang T, Bao Y, Hao H. An overview of nanomaterials for water and wastewater treatment. Advances in Materials Science and Engineering. 2016;**2016**: 1-10. DOI: 10.1155/2016/4964828

[58] Xu P, Zeng GM, Huang DL, Feng CL, Hu S, Zhao MH, et al. Use of

iron oxide nanomaterials in wastewater treatment: A review. Science of the Total Environment. 2012;**424**:1-10. DOI: 10.1016/j.scitotenv.2012.02.023

[59] Phuengprasop T, Sittiwong J, Unob F. Removal of heavy metal ions by iron oxide coated sewage sludge. Journal of Hazardous Materials. 2011;**186**:502-507. DOI: 10.1016/j.jhazmat.2010.11.065

[60] Marchetti SG, Fraga MA, Paulino PN, de Oliva ST, Borges SMS, da Silva LA, et al. Methylene blue oxidation over iron oxide supported on activated carbon derived from peanut hulls. Catalysis Today. 2016;**289**:237-248. DOI: 10.1016/j.cattod.2016.11.036

[61] Villalobos M, Bargar J, Sposito G. Mechanisms of Pb(II) sorption on a biogenic manganese oxide. Environmental Science & Technology. 2005;**39**:569-576. DOI: 10.1021/es049434a

[62] Yang X, Makita Y, Liu ZH, Sakane K, Ooi K. Structural characterization of self-assembled MnO_2 nanosheets from birnessite manganese oxide single crystals. Chemistry of Materials. 2004;**16**:5581-5588. DOI: 10.1021/cm049025d

[63] Zhang P, Wu SS. Synthesis and characterization of poly(N-isopropylacrylamide)-modified zinc oxide nanoparticles. Advances in Materials Research. 2013;**771**:141-146. DOI: 10.4028/www.scientific.net/amr.771.141

[64] Anjum M, Miandad R, Waqas M, Gehany F, Barakat MA. Remediation of wastewater using various nano-materials. Arabian Journal of Chemistry. 2016:7. DOI: 10.1016/j.arabjc.2016.10.004

[65] Ohashi N, Grasset F, Duguet E, Mornet S, Park D, Roisnel T, et al. Surface modification of zinc oxide nanoparticles by aminopropyltriethoxysilane. Journal of Alloys and Compounds. 2003;**360**:298-311. DOI: 10.1016/s0925-8388(03)00371-2

[66] Gao Z, Wei L, Yan T, Zhou M. Modification of surface layer of magnesium oxide via partial dissolution and re-growth of crystallites. Applied Surface Science. 2011;**257**:3412-3416. DOI: 10.1016/j.apsusc.2010.11.035

[67] Pilarska AA, Klapiszewski Ł, Jesionowski T. Recent development in the synthesis, modification and application of $Mg(OH)_2$ and MgO: A review. Powder Technology. 2017;**319**:373-407. DOI: 10.1016/j.powtec.2017.07.009

[68] Cai L, Chen J, Liu Z, Wang H, Yang H, Ding W. Magnesium oxide nanoparticles: Effective agricultural antibacterial agent against *Ralstonia solanacearum*. Frontiers in Microbiology. 2018;**9**:1-19. DOI: 10.3389/fmicb.2018.00790

[69] Qu S, Huang F, Yu S, Chen G, Kong J. Magnetic removal of dyes from aqueous solution using multi-walled carbon nanotubes filled with Fe_2O_3 particles. Journal of Hazardous Materials. 2008;**160**:643-647. DOI: 10.1016/j.jhazmat.2008.03.037

[70] Ai L, Zhang C, Liao F, Wang Y, Li M, Meng L, et al. Removal of methylene blue from aqueous solution with magnetite loaded multi-wall carbon nanotube: Kinetic, isotherm and mechanism analysis. Journal of Hazardous Materials. 2011;**198**:282-290. DOI: 10.1016/j.jhazmat.2011.10.041

Section 2

Biological Properties of Nanoparticles

Chapter 6

Cellular and Molecular Impact of Green Synthesized Silver Nanoparticles

Paritosh Patel, Puja Kumari, Suresh K. Verma and M. Anwar Mallick

Abstract

Toxicity and biocompatibility of silver nanoparticles are of a major concern due to their extensive production regardless of their application in current industries. Information about toxicology or biocompatibility is crucial regarding their proper utilization and application in clinical as well as environmental aspect. This chapter describes in detail about the different techniques and technology of synthesis of silver nanoparticles and explains their different physiochemical properties in context of the current research scenario. Further, it also explains the biocompatibility and toxicity of silver nanoparticles at cellular and molecular aspects. The mechanism of their toxicity has been described keeping in view of the recent research done. In brief, it reveals detail knowledge of the cellular and molecular impact of silver nanoparticles.

Keywords: silver nanoparticles, toxicology, oxidative stress, apoptosis

1. Introduction

Really revolutionary nanotech items, materials and application for example nanorobotics, are years long in the future. But what qualifies as "Nanotechnology" today is fundamental innovation that is going on in research centers everywhere throughout the world. Products of Nanotech which are on business sector today are generally steadily improved products (utilizing evolutionary nanotechnology) where some types of Nano-empowered materials (for example, Carbon nanotubes, nanocomposite structure of nanoparticles of specific substance) or nanotech process (for example Nano-patterning or Quantum Dots for medicinal imaging) is utilized in the assembling procedure. In their progressing and ongoing journey to improve existing products by making smaller parts and better execution materials, all at a lower cost, the number of organization that will make "Nano products" will become extremely fast and soon make up the most of all organization across numerous businesses.

Nanomaterials (NMs) have picked up noticeable quality in technological progressions due to their tunable synthetic, physical and organic properties with improved execution over their bulk counter partners. They are arranged depending on their origin, size, shape and composition. The capacity to anticipate the remarkable properties of NMs expand the estimation of each classification. Nanomaterials speak to an active/functioning area of research and techno-economic parts in numerous application areas. NMs are depicted as a material with a length of 1–1000 nm in at least one dimension [1]. In any case, a single globally acknowledge definition for NMs does not exist. The diverse association has a distinction in assessment in defining NMs. As indicated by the Environmental Protection Agency (EPA), NMs can display remarkable properties unique than the equal chemical compound in a bigger dimension [2]. The US Food and Administration (USFDA) likewise alludes to NMs as "materials that have at least one dimension dependent phenomena" [2]. The International Organization for Standardization (ISO) has depicted NMs as a "Materials with any external nanoscale measurement or having internal nanoscale surface structure" [2]. As of late, the British Standard Institution proposed the following definition for the scientific terms that have been utilized:

- *Nanoscale*: Approximately 1–1000 nm size range [1, 3].

- *Nanoscience*: The science and investigation of matter at the nanoscale that manages to understand their size and structure-dependent properties and compares at the rise of individual atoms or molecules or bulk materials related differences [1, 3].

- *Nanotechnology*: manipulation and control of matter on a nanoscale measurement by utilizing scientific logical knowledge of different industrial and biomedical applications [1, 3].

- *Nanomaterials*: Materials with any inside or outside structure on the nanoscale measurements [1, 3].

- *Nano-objects*: Materials that have at least one or more peripheral nanoscale measurements.

- *Nanoparticles*: Nano-objects with three outer nanoscale measurements. The terms Nano rod or Nano plate are utilized, rather than nanoparticles (NP) when the longest and the shortest axes length of a nano-object are unique [1, 3].

- *Nanofiber*: When two comparable exterior nanoscale measurements and a third measurement are available in a nanomaterial, it is alluded to as a nanofiber [1, 3].

- *Nanocomposite*: Multiphase structure with at least one phase on the nanoscale measurement [1, 3].

- *Nanostructure*: Composition of interconnected parts in the nanoscale area [1, 3].

- *Nanostructured materials*: Materials containing interior or surface nanostructure [1, 3].

The nanoparticles shows remarkable chemical, physical and natural properties at nanoscale contrasted with their respective particles at higher scales. This phenomenon is because of a moderately bigger surface region to the volume, expand reactivity or stability in a synthetic procedure, improved mechanical strength and so forth. These properties of nanoparticles have prompted its utilization of different applications [3]. Nanoparticles have been utilized in medication (drug delivery), in food industries, gene delivery and Cancer therapy and so on [3]. The nanoparticles are of various size, structure and shape. It well may be tubular, conical, spherical, hollow core, cylindrical, spiral, flat and so forth or sporadic and contrast from 1 to 100 nm in size. Nanomaterials/or nanoparticles are utilized in an expansive range of use. Today they contained in numerous products and utilized in different technologies. Most Nano items created on an industrial scale are nanoparticle, in spite of the fact that they likewise emerge as by-products in the manufacturing of other materials [4, 5]. Explicit synthesis is utilized to create the different nanoparticles, coating composite and dispersion. Characterized production and reaction condition is pivotal in acquiring such size-dependent molecule. Particle shape, crystallinity, chemical composition and size can be constrained by pH- value, synthetic arrangement (chemical), temperature, procedure control and surface modification [5].

Two fundamental procedures are utilized to create nanoparticles: "Top-down" and "Bottom-up". The expression "Top-down" alludes here to the mechanical squashing of source materials utilizing a milling procedure. In the "Bottom-up" strategy, structures are developed by the synthetic procedure. The determination of the individual procedure relies upon the compound organization and the desire features indicated for the nanoparticles [6] (**Figure 1**) (**Table 1**).

Strangely, the morphological parameters of NPs can be tweaked by shifting the chemical concentration and reaction condition for example pH and temperature. However, if these synthesized NMs are exposed to the real application, then they can experience the following impediment, which is stability in a threatening situation, absence of comprehension in fundamental mechanism and modeling factors, bioaccumulation or toxicity quality, extensive examination requirements recycle, reuse, regeneration. In true word, it is desirable that the properties, behavior and types of nanomaterials ought to be improved to meet the aforementioned points. Then again, these impediments are opening new and extraordinary opportunities in this developing field of research.

To counter those restrictions a new era of green synthesis methodologies is increasing incredible in recent research and innovative work on material science and technologies. Essentially green synthesis will straightforwardly help uplift the ecological friendliness as they are generated through clean up, regulation/guideline, control and remediation process additionally there are few parts like the decrease of derivatives, decrease of contamination, prevention and minimalization of waste

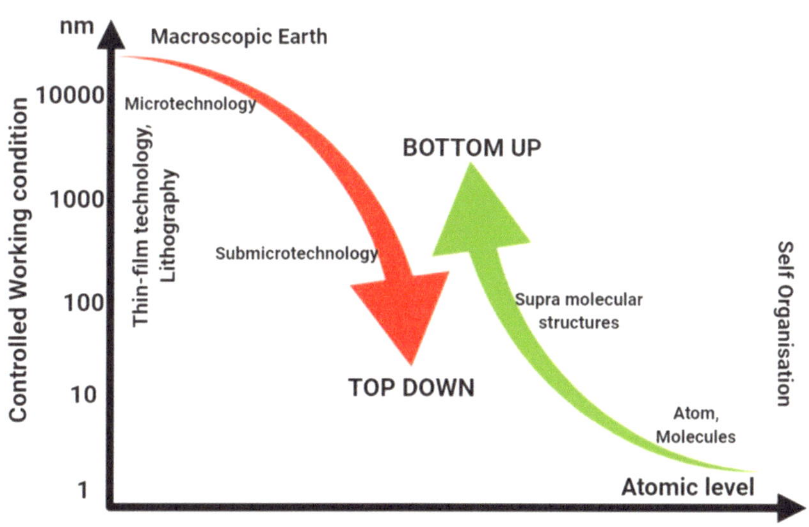

Figure 1.
Methods of nanoparticles production: top-down and bottom-up (image: Laboratory for micro and nanotechnology, Paul Scherrer Institut).

Category	Method	Nanoparticles
Bottom-up	Pyrolysis	Carbon and metal oxide based
	Biosynthesis	Organic polymers and metal based
	Spinning	Organic polymers
	Sol-gel	Carbon metal and metal oxide based
	Chemical vapor deposition (CVD)	Carbon and metal based
Top-down	Sputtering	Metal based
	Laser ablation	Carbon based and metal oxide based
	Thermal decomposition	Carbon and metal oxide based
	Nanolithography	Metal based
	Mechanical milling	Metal, oxide and metal oxide based

Table 1.
Categories of the nanoparticles synthesized from the various methods [1].

and ultimately the utilization of more secure solvent during synthesis process as well as renewable stock. Green synthesis is required to stay away from the production of undesirable or unsafe products through the build-up of reliable, maintainable and eco-friendly methods. Green synthesis of metallic nanoparticles has been embraced to suit different organic material (for example, algae, bacteria, plant extract and fungi) (**Figure 2**) [7]. Among the accessible green methods of synthesis for metal and metal oxide NPs, usage of plant extract is a fairly straightforward and simple procedure to create nanoparticles at large scale with respect to fungi and bacteria mediated synthesis. Synthesis of metal and metal oxide NPs, plant biodiversity has been comprehensively considered to be because of the availability of effective phytochemicals in different plant extract, particular in leaves such as amide, flavones, phenols, terpenoids, ketones, ascorbic acid, aldehyde and carboxylic acids. These components are equipped of reducing metal salts into metal NPs [7, 8] (**Tables 2** and **3**).

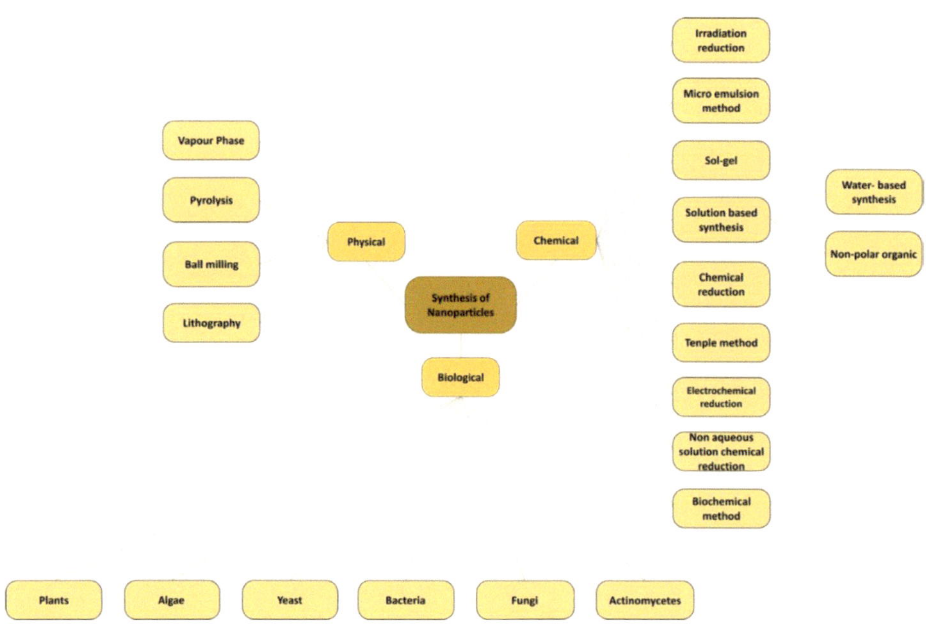

Figure 2.
Different methods for the synthesis of nanoparticles [4, 7].

Sr. no.	Species	Nanoparticles	Size (nm)	Morphology	Application
1	*Lactobacillus casei*	Silver	20–50	Spherical	Drug delivery, bio-labeling
2	*Desulfovibrio desulfuricans*	Gold	20–50	Spherical	Catalysis
3	*E. coli*	Cadmium	2–5	Fluorescent labels	Wurtzite structures
4	*Klebsiella pneumonia*	Silver	28–122	Spherical	Optical receptor, antimicrobial
5	*Aquaspirillum magnetotacticum*	Iron oxide	40–50	Octahedral prism	—
6	*Coriolus versicolor*	Silver	25–75	Spherical	Water-soluble metallic catalyst
7	*Penicillium brevicompactum*	Silver	23–105	Crystalline spherical	Antimicrobial agent
8	*Phoma glomerata*	Silver	60–80	Spherical	Antimicrobial agent
9	*Saccharomyces cerevisiae* broth	Gold, silver	4–15	Spherical	Catalysis
10	*Aspergillus flavus* TFR7	Titanium dioxide	12–15	Spherical	Plant nutrient fertilizer

Table 2.
Synthesis of metallic nanoparticles from various biological species (bacteria) [7].

Sr. no.	Species	Nanoparticles	Size (nm)	Morphology	Application
1	*Eucalyptus citriodora* (neelagiri)	Silver	20	Spherical	Antibacterial
2	*Cymbopogon flexuosus* (lemon grass)	Gold	200–500	Spherical, triangular	Infrared-absorbing optical coating
3	*Syzygium aromaticum* (clove buds)	Gold	5–100	Irregular	Detection and destruction of cancer cells
4	*Mentha piperita* (peppermint)	Silver	5–30	Spherical	Kill microbes
5	*Medicago sativa* (alfalfa)	Gold	2–40	Hexagonal, tetrahedral, icosahedral	Labeling in structural biology, paints
6	*Morus* (mulberry)	Silver	15–20	Spherical	Antimicrobial activity
7	*Aloe barbadensis* Miller(*Aloe vera*)	Gold, silver	10–30	Spherical, triangular	Cancer hyperthermia, optical coating
8	*Coriandrum sativum* (coriander)	Gold	6.75–57.91	Spherical, triangular	Drug delivery, tumor imaging
9	*Azadirachta indica* (neem)	Gold, silver	5–35	Spherical, hexagonal	Remediation of toxic metal
10	*Terminalia catappa* (almond)	Gold	10–35	Spherical	Biomedical field

Table 3.
Synthesis of metallic nanoparticles from various plant extract [7].

2. Classification of nanoparticles

2.1 Organic nanoparticles

Dendrimers, micelles, liposomes and ferritin are usually known as natural nanoparticles or polymers. These nanoparticles are biodegradable, non-toxic and a few particles for example, micelles and liposomes have a hollow center otherwise known as nanocapsules [9].

2.2 Inorganic nanoparticles

Inorganic nanoparticles are not comprised of carbon. Metal and metal oxide based nanoparticles are commonly classified as inorganic nanoparticles.

- *Metal based*: Nanoparticles that are integrated from metals to Nano size either by ruinous or constructive strategies are metal based nanoparticles [1, 9]. Practically every one of metal can be synthesized into their nanoparticles. The normally utilized metals for nanoparticles are aluminum, cobalt, gold, silver, zinc, iron, copper, and cobalt [1, 9]. Nanoparticles have a distinctive size extends from 10 nm to 100 nm.

- *Metal oxides based*: The metal oxides based nanoparticles are orchestrated to adjust the properties of their respective metal based nanoparticles.

- *Cerium oxide*: These nanoparticles have excellent properties when contrasted with their metal partner. For example, zinc oxide. Iron oxide, silicon dioxide, magnetite, etc.

2.3 Carbon based

The nanoparticles made totally of carbon are known as carbon-based [1, 9, 10]. They can be classified as:

- *Fullerenes*: Fullerenes is a carbon particle that is spherical on shape and made up of carbon molecules held together by sp^2 hybridization. Around 28–1500 carbon atoms form the spherical structure with diameter of 8.2 nm for a single layer and for a multi-layered fullerenes 4–36 nm [9, 10].

- *Graphene*: Graphene is an allotrope of carbon. It is a hexagonal system of honeycomb lattice made of carbon atoms in a 2-D planar surface. The thickness of the graphene is of 1 nm [9, 10].

- *Carbon nano tubes*: In this, nano foil which has a honeycomb lattice of carbon atoms is twisted into a hollow cylinder to frame nanotubes of measurements as low as 0.7 nm [1, 10].

- *Carbon nanofiber*: When graphene nano foil used to produce carbon nanofiber as carbon nanotubes however twisted into a cone or cup shape than a regular cylindrical tube [1, 10].

- *Carbon black*: It is an undefined material comprised of carbon, generally spherical in shape with diameter measurements up to 20–70 nm. Interaction between the particles is high to such as an extent that they aggregate and the agglomeration are seen as of 500 nm [1, 10].

3. Nanoparticle as a threat

Nano-technology has acquired an incredible revolution in the industrial division. Due to their exceptional physiochemical and electrical properties, Nano-sized materials have increased a great deal of fascination in the field of hardware, biotechnology and aeronautic design. It is additionally being utilized in the field of medicine NPs similar to the novel delivery system for drugs, DNA and so on. Human is exposed to different non-scale materials since the new developing field of nanotechnology has turned into another danger to human life [11, 12]. The proposed hypothesis is that the NPs of size under 10 nm act similar to gas and can enter human tissues effectively and may abrogate the cell typical biochemical condition [11, 13]. There have been studies on human and murine models that the NPs are exposed through orally they are circulated to the spleen, liver, heart

Figure 3.
The following figure represents usages of nanotechnology/nanoparticles in different field [6].

and lungs also to the brain and gastrointestinal zone, some other exposure routes may incorporate skin, ingestion, inhalation and injection. Some designed NPs are being utilized in many products with direct exposure to people, for instance, ZnO NPs are added to numerous items including cotton texture, Food packaging and rubber for its freshening up and antibacterial attributes, TiO_2 NPs are utilized in food coloring, makeup, skincare item and tattoo pigments, Fe_2O_3 NPs utilized in the final polish on metallic gems (jewelries) [12]. It has been seen that life expectancy of the nanoparticles in human is around 700 days in which it reliably has a risk to the body. Nanoparticles have an incredible risk to human's wellbeing when contrasted with large-sized particles of the similar chemical compound and it is commonly said that toxicities are contrarily corresponding to the size of the nanoparticles [14, 15]. As the utilization of engineered nanoparticles keeps on developing exponentially, an unintended and intended exposure may happen, which will prompt a high level of human wellbeing hazard. End product users, occupationally exposed subjects and the overall population may be in danger of antagonistic impact (**Figure 3**).

The physiochemical properties of NPs impact how they interact with cells and thus, their potential danger. Studies have demonstrated the different properties that make some nanoparticles more toxic than others. Hypothetically, molecules size is likely to add to cytotoxicity. Smaller NPs have a bigger specific surface area and thus in this way increasingly accessible surface area to interact with cell components for example, carbohydrates, protein, nucleic acids and fatty acids. Nanoparticles with small size are liable to enter the cells, causing cellular damage. Some nanoparticles lethality were seen as a function of both size and specific surface area. It has additionally been seen that size of NPs has seen to correspond with reactive oxygen species (ROS) generation when comparing the amount of ROS generation per surface area within certain size range [14–16]. Nanoparticles size between 10 nm or > 30 nm creates comparable level of ROS per surface area. In any case, there was a sensational increment in ROS production per unit surface in particle expanding from 10 to 30 nm. This information or data disclose to us the bits of knowledge with respect to the perplexing connection between NPs properties and Nano toxicity.

4. Toxicity of silver nanoparticles

Silver nanoparticles are progressively utilized in different fields, including health care, medical, food, consumer and industrial purpose because of their novel physical and chemical properties. Because of their unconventional properties, they have been utilized for a few applications, as in medical device coating, drug delivery, health care products, and food industry, as anticancer agent and orthopedics and also as anti-bacterial agents. AgNPs by a long shot the most generally utilized in customers items, for example, in kitchen utensils, toothpaste, bedding, deodorants, nursing bottles, washing machines, nipples and humidifiers [17]. So as to satisfy the necessity of silver NPs different strategy have been utilized for synthesis, conventional technique like chemical and physical strategies have been utilized, yet they are by all accounts expensive and toxic/hazardous [18].

An organic methodology has been utilized in the synthesis of AgNPs utilizing microorganisms, fungi and plant extract prompting to reliable alterative to chemical and physical techniques in acquiring the nanoparticles in controlled particle size. It has been seen that green synthesis of AgNPs with various stabilizing agents, for example, polyethylene glycol, alcohol vinyl, dextran, cyclodextrins and utilizing

DOI: http://dx.doi.org/10.5772/intechopen.90717

"andeli" (rose extract) [19–22]. This expands utilization of AgNPs in different materials has prompted a more straightforward and direct exposure in human and raised potential dangers to health issues. In an in vivo examination (Sprague-Dawley rodents) they were dealt with orally for 28 days with AgNPs in spite of the fact that there was no observable difference in clinical sign and neither in any difference in body or organ weight. The impacts on blood biochemistry have been seen to increase in cholesterol at high doses of AgNP which indicates that hepatotoxicity and increment in alkaline phosphatase. In another investigation, F344 rodents were fed AgNPs for a time period of 90 days and a decrease in body weight in males was seen following a month of exposure and a dose-dependent change was seen in cholesterolemia and alkaline phosphatases activity which proposed that 125 mg/kg body weight of AgNPs may cause liver harm (**Figure 4**).

It is commonly realized that NPs can be absorbed by the digestive tract not just through the M-cells in the Peyer's patch yet additionally by numerous organs as shown in **Figure 5**. Culture of human mesenchymal stem cells incubated with 0.1 $\mu g\ ml^{-1}$ of normal human lungs fibroblast cells or albumin-capped silver nanoparticles or human glioblastoma cells starch incubated with capped silver nanoparticle (AgNPs) showed genotoxicity up to dosages of 50 $\mu g\ ml^{-1}$. Albumin-capped AgNPs has been demonstrated to be more genotoxic than polysaccharide-capped AgNPs. Silver nanoparticles exposed to in vivo models like mice were seen to be more toxic than fish to capped AgNPs. However NPs toxicity/genotoxicity was seen as more when the concentration of albumin-capped NPs was increased to >100 $\mu g\ ml^{-1}$ (**Table 4**).

In recent investigation, it has been seen that hepatotoxicity, pulmonary inflammation, genotoxicity, neurotoxicity, inflammatory effects and cytotoxicity have been related with various shape and size of silver nanoparticles. A large number of articles have been proposed that silver nanoparticles assume a noteworthy job in prompting Reactive Oxygen Species (ROS) which in returns lead to cell cytotoxicity and genotoxicity. It has been seen that cytotoxicity has been closely identified to the generation of ROS. For example interaction of AgNPs with mitochondrial can be seen that: 95% of the cell's energy is generated by mitochondria as it is the powerhouse of the cells. It's a significant and essential piece of the cell. ROS generation is possible because of the superoxide spillage through the membrane. The interaction of AgNPs with mitochondria which prompts generation of ROS can be clarified in a manner when Ag^+ and silver nanoparticles have a high affinity (–SH) thiol group in cysteine residues. AgNPs disrupts the membrane proteins integrity of the mitochondria, also hampers the membrane permeability of the membrane and abrogate the mitochondrial functions (**Figure 6**).

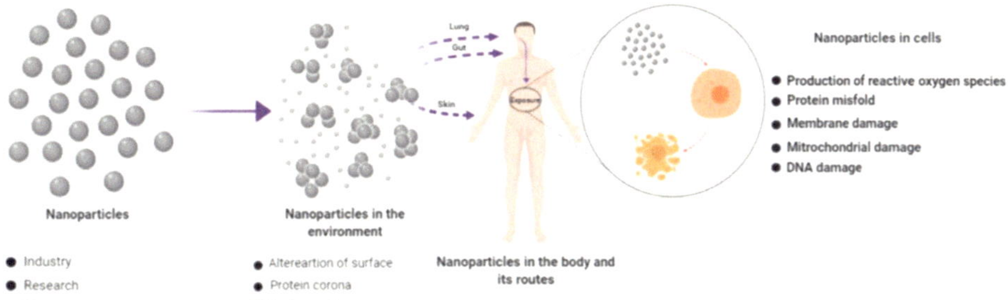

Figure 4.
Nanoparticles pathway and toxicological impact [13, 14].

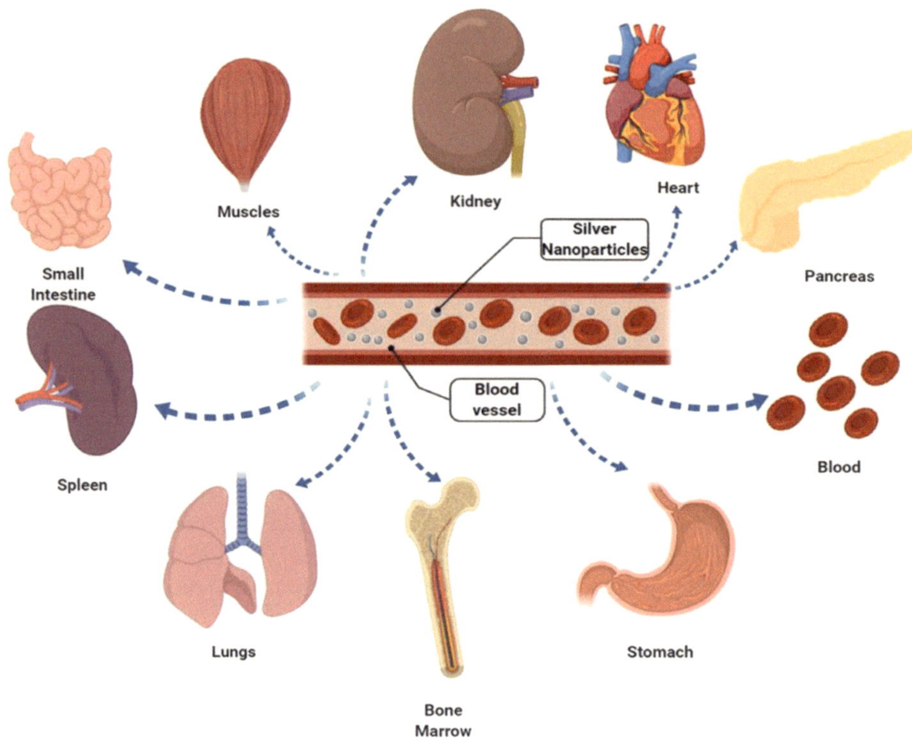

Figure 5.
Main target organ of silver nanoparticles.

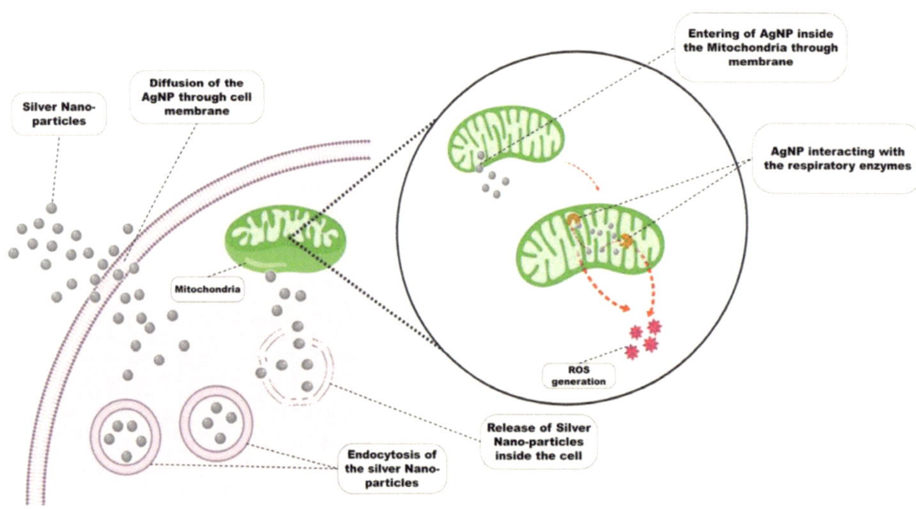

Figure 6.
This figure represents the endogenous ROS production, which are involved in oxidative stress. Interaction between mitochondrial and AgNP generate ROS from mitochondria which leads to cell death [23].

A great deal of studies has demonstrated a connection in ROS generation by silver nanoparticles, oxidative stress and cytotoxicity. Numerous toxicological changes have been reported in embryos when they are exposed to nanoparticles, for example, changes in oxidative stress markers such as apoptosis, changes in expression of genes and lipid oxidation, etc. It has been seen that massive production of free radicals lead to the generation of pro-inflammatory cytokines and furthermore initiation of NOX/

DOI: http://dx.doi.org/10.5772/intechopen.90717

Sr. no.	Classification	Cell	Size (nm)	Cytotoxicity	Genotoxicity
1	Vertebrate/chordata/ mammalian	IMR-90 Human lung fibroblast	6–20	50 μg ml^{-1}	Comet assay 5 μm (50 comets were analyzed per concentration) (DNA damage at 50 μg ml^{-1})
2	Vertebrate/chordata/ mammalian, human	A549 Human lung carc.	78	12.5 μg ml^{-1}	In this it was indicated that silver NPs mediated ROS-induced genotoxicity
3	Vertebrate/chordata/ mammalian	mES Mouse embryonic stem cells	25	50 μg ml^{-1}	50 μg ml^{-1} (this conc. upregulates the DNA damage repair proteins Rad51 and phosphorylated-H2AX expression and even upregulates cell cycle checkpoint protein p53)
4	Vertebrate/chordata/fish	OLHN12 Medaka fish	20–30	1.3 μg ml^{-1}	1.2 μg ml^{-1} aneuplidy 15.8%
5	Vertebrate/chordata/ mammalian	BRL 3A Rat liver cells	25	50 μg ml^{-1}	10 μg ml^{-1}
6	Plantae/liliopsida	*Allium cepa* 5000 cells	24–55	ROS-induced up to 10 μg ml^{-1}cell death and DNA damage doses 20 μg ml^{-1}	No genotoxicity
7	Vertebrate/chordata/fish	Primary trout hepatocytes N = 3, each treatement	35	Significant toxicity from 500–1000 μg ml^{-1}	Didn't cause significant lactate dehydrogenase release
8	Vertebrate/chordata/ mammalian	MEF Mouse embryonic fibroblasts	25	50 μg ml^{-1}	50 μg ml^{-1} this concentration upregulates the Rad51 and phosphorylates-H2AX expression and also upregulates p53
9	Vertebrates/chordata/ mammalian, human	Human lymphocytes	25–45	Till 400 μg ml^{-1}	Till 50 μg ml^{-1} no DNA damage
10	Vertebrate/chordata/ mammalian, human	HepG2 Human hepatocytes 150 cell per group	15–20	13 μg ml^{-1} (58% survival)	Here it was indicated that the AGNPs induces ROS-induced nontoxicity

Table 4.
In vivo cytotoxicity and genotoxicity effects of AgNPs in different organisms [25].

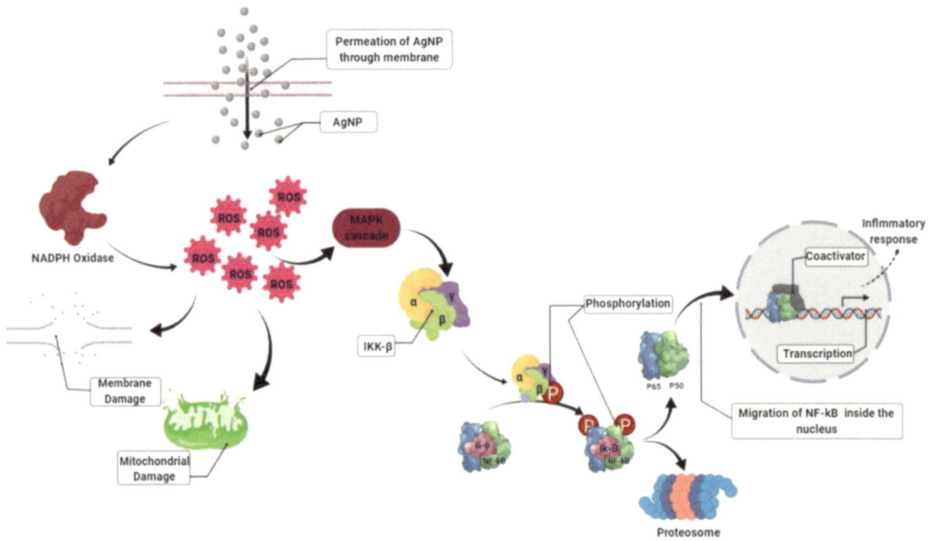

Figure 7.
Pictorial representation of activation of cellular mechanisms of inflammatory signal when exposed to AgNP generated from ROS and by strengthening of NADPH oxidase activity. By MAP kinases pathway, activation of oxidative IKK-B which is induced by stress leads to NF-kB translocation and expression of marker and potential mediators of inflammation increases, mitochondrial damage and membrane damage which can cause toxicity in cell and leads to death by apoptosis [24].

NADPH oxidase family. It must be noticed that other than inflammatory effects of oxygen radicals, oxidants helps the release of inflammatory mediators by activating transcription factor including AP-1, hypoxia inducing factor and NF-kB and which prompts to oxidative stress and inflammation. Oxidative stress might have a double role, first as "effector" (by oxidant discharge and induced toxicity) and secondly as "modulator" (managing transcription factor) of chronic micro-inflammation process. This interaction among inflammation and oxidative stress in an amplified manner may prompt to the deleterious impacts brought by silver nanoparticles and which can prompt to DNA damage and cell demise by apoptosis (**Figure 7**).

5. Conclusion

In brief, the ingenious and extensive demand of nanoparticles has led to their extensive production. After use modalities path their way towards exposure to environment as well as to the human health moreover the product which are being used are also in direct contact with the human tissue. Prolong accumulation of the nanoparticle particularly talking about silver nanoparticles.

Author details

Paritosh Patel[1], Puja Kumari[2], Suresh K. Verma[1] and M. Anwar Mallick[2]*

1 School of Biotechnology, KIIT University, Bhubaneswar, India

2 Advance Science and Technology Research Centre, Vinoba Bhave University, Hazaribagh, India

*Address all correspondence to: amallick1@rediffmail.com

References

[1] Ealias AM, Saravanakumar MP. A review on the classification, characterisation, synthesis of nanoparticles and their application. IOP Conference Series: Materials Science and Engineering. 2017;**263**:032019. DOI: 10.1088/1757-899X/263/3/032019

[2] Jeevanandam J, Barhoum A, Chan YS, Dufresne A, Danquah MK. Review on nanoparticles and nanostructured materials: History, sources, toxicity and regulations. Beilstein Journal of Nanotechnology. 3 Apr 2018;**9**:1050-1074. DOI: 10.3762/bjnano.9.98

[3] Heera P, Shanmugam S. Review article nanoparticle characterization and application: An overview. 2015;**4**:379-386

[4] Christian P, Baalousha M, Hofmann T. Nanoparticles: Structure, properties, preparation and behaviour in nanoparticles: Structure, properties, preparation and behaviour in environmental media. Ecotoxicology. Jul 2008;**17**(5):326-343. DOI: 10.1007/s10646-008-0213-1

[5] Prakash Sharma V, Sharma U, Chattopadhyay M, Shukla VN. Advance applications of nanomaterials: A review. Materials Today: Proceedings. 2018;**5**:6376-6380. DOI: 10.1016/j.matpr.2017.12.248

[6] Zielonka A, Klimek-ochab M. Fungal synthesis of size-defined nanoparticles. Advances in Natural Sciences: Nanoscience and Nanotechnology. 2017;**8**:043001. DOI: 10.1088/2043-6254/aa84d4

[7] Singh J, Dutta T, Kim KH, Rawat M, Samddar P, Kumar P. Green synthesis of metals and their oxide nanoparticles: Applications for environmental remediation. Journal of Nanobiotechnology. 2018;**16**:1-24. DOI: 10.1186/s12951-018-0408-4

[8] Kumari P, Panda PK, Jha E, Kumari K, Nisha K. Mechanistic insight to ROS and Apoptosis regulated cytotoxicity inferred by green synthesized CuO nanoparticles from *Calotropis gigantea* to embryonic Zebrafish. Scientific Reports. 2017: 1-17. DOI: 10.1038/s41598-017-16581-1. Article number: 16284

[9] Grose A, Grose A. What are the different types of therapy? In: Are You Considering Therapy. 2018. pp. 1-96. DOI: 10.4324/9780429471940-1

[10] Narendra Kumar SK. Essentials in nanoscience and nanotechnology. Carbon-Based Nanomaterials. John Wiley & Sons, Inc; 16 Mar 2016. Print ISBN: 9781119096115, Online ISBN: 9781119096122. DOI: 10.1002/9781119096122. Chapter 5

[11] Huang YW, Cambre M, Lee HJ. The toxicity of nanoparticles depends on multiple molecular and physicochemical mechanisms. International Journal of Molecular Sciences. Dec 2017;**18**(12):2702. DOI: 10.3390/ijms18122702

[12] Bahadar H, Maqbool F, Niaz K, Abdollahi M. Toxicity of nanoparticles and an overview of current experimental models. Iranian Biomedical Journal. 2016;**20**:1-11. DOI: 10.7508/ibj.2016.01.001

[13] De Matteis V, Rinaldi R. Toxicity assessment in the nanoparticle era. Advances in Experimental Medicine and Biology. 2018;**1048**:1-19. DOI: 10.1007/978-3-319-72041-8_1

[14] Sukhanova A, Bozrova S, Sokolov P, Berestovoy M, Karaulov A, Nabiev I. Dependence of nanoparticle toxicity on their physical and chemical properties. Nanoscale Research Letters. 7 Feb 2018;**13**(1):44. DOI: 10.1186/s11671-018-2457-x

[15] Gálvez Pérez V, Gózalo CT, Leite PEC, Pereira MR, Granjeiro JM, Alkilany AM, et al. Toxicología de las nanopartículas. Revista Complutense de Ciencias Veterinarias. 2015;**33**:2313-2333. DOI: 10.1016/j.addr.2011.09.001

[16] Liu R, Rallo R, George S, Ji Z, Nair S, Nel AE, et al. Classification NanoSAR development for cytotoxicity of metal oxide nanoparticles. Small. 2011;**7**:1118-1126. DOI: 10.1002/smll.201002366

[17] Zhang XF, Liu ZG, Shen W, Gurunathan S. Silver nanoparticles: Synthesis, characterization, properties, applications, and therapeutic approaches. International Journal of Molecular Sciences. 2016;**17**. DOI: 10.3390/ijms17091534

[18] Akter M, Sikder MT, Rahman MM, Ullah AKMA, Hossain KFB, Banik S, et al. A systematic review on silver nanoparticles-induced cytotoxicity: Physicochemical properties and perspectives. Journal of Advanced Research. 2018;**9**:1-16. DOI: 10.1016/j.jare.2017.10.008

[19] Erdogan O, Abbak M, Demirbolat GM, Birtekocak F, Aksel M, Pasa S, et al. Green synthesis of silver nanoparticles via Cynara scolymus leaf extracts: The characterization, anticancer potential with photodynamic therapy in MCF7 cells. PLoS One. 2019;**14**:1-15. DOI: 10.1371/journal.pone.0216496

[20] Wu X, Lu C, Zhou Z, Yuan G, Xiong R, Zhang X. Green synthesis and formation mechanism of cellulose nanocrystal-supported gold nanoparticles with enhanced catalytic performance. Environmental Science. Nano. 2014;**1**:71-79. DOI: 10.1039/c3en00066d

[21] Roy A, Bulut O, Some S, Mandal AK, Yilmaz MD. Green synthesis of silver nanoparticles: Biomolecule-nanoparticle organizations targeting antimicrobial activity. RSC Advances. 2019;**9**:2673-2702. DOI: 10.1039/c8ra08982e

[22] Arshad A. Bacterial synthesis and applications of nanoparticles. Journal of Nanoscience and Nanotechnology. 2017;**11**:119

[23] Silver nanoparticles: Electron transfer, reactive oxygen species, oxidative stress, beneficial and toxicological effects. Mini Review. DOI: 10.1002/jat.3654

[24] Silver nanoparticles: Their potential toxic effects after oral exposure and underlying mechanisms—A review. DOI: 10.1016/j.fct.2014.12.019

[25] Silver nanoparticles: A brief review of cytotoxicity and genotoxicity of chemically and biogenically synthesized nanoparticles. DOI: 10.1002/jat.2780

Chapter 7

Theranostic Nanoparticles and Their Spectrum in Cancer

Anca Onaciu, Ancuta Jurj, Cristian Moldovan and Ioana Berindan-Neagoe

Abstract

Nanoparticles offer a lot of advantageous backgrounds for many applications due to their physical, chemical and biological properties. Their different composition (metals, lipids, polymers, peptides) and shapes (spheres, rods, pyramids, flowers and so on) are influenced by the synthesis methods and functionalization procedures. However, in the medical field, researchers focus on the biocompatibility and biodegradability of the nanoparticles in their attempts for a targeted therapy in which the nanocarriers need to bypass certain biological barriers. Moreover, the increased interest in molecular imaging has brought nanoparticles in the spotlight for their applications in two distinct directions: therapy and diagnosis. Furthermore, recent advances in nanoparticle designs have introduced novel nano-objects suitable as both detection and delivery systems at the same time, thus providing theranostic applications.

Keywords: nanoparticles, nano-oncology, targeted therapy, molecular imaging, diagnosis

1. Introduction

Nanomedicine is able to study the organism and especially the disease at the nanoscale level and offers a lot of structural and functional information for the development of new therapeutics and diagnosis strategies [1]. Nano-oncology refers to the applications of nanotechnology in the oncology medical field.

Oncological malignancies affect worldwide population with an incidence of 18.1 million new cancer cases and 9.6 million cancer deaths (GLOBOCAN 2018). Usually, the most used treatment scheme is surgery, radiotherapy and chemotherapy. These strategies are not very efficient because it does not only affect the disease site, but healthy tissues too, and in many cases, cancer can develop therapy resistance [2].

Nanotechnology tools have potential to overcome the side effects and the inefficiency of some therapies. Due to its small size, nanoparticles (NPs) can be used for molecular characterization of the disease, and based on this, it can contribute to discover new therapies. Moreover, various oncological chemotherapeutics are nanoformulated and now are involved in clinical trials [3].

Besides drug encapsulation, NPs can be used for the delivery of growth factors and other compounds applied in tissue engineering. On the other hand, NPs' properties are advantageous for new sensing and molecular imaging tools development (**Figure 1**).

 IntechOpen

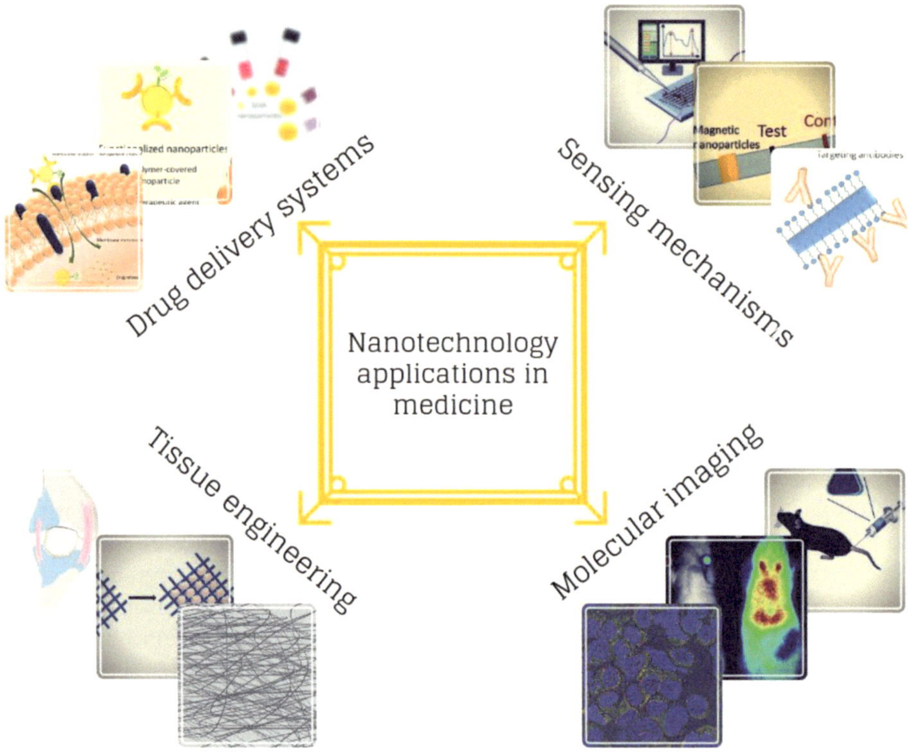

Figure 1.
Nanotechnology applications in medicine.

For each of these applications, NPs' formulations involve various encapsulation procedures, which need to meet specific characteristics. Firstly, the NPs should not interfere with the encapsulated compound pharmacological activity, and it has to prevent its premature degradation and to become biodegradable at the tumor site, thus decreasing its toxicity [4]. Secondly, for sensing applications, the nanosystem needs to have some unique chemical, electrical, and catalytical properties to provide accuracy of the measurements [5]. On the other hand, for molecular imaging applications, the NPs benefit from their optical properties like fluorescence in various spectra. Also, the features such as biocompatibility, stability and long circulation time are very important [6–8].

Theranostic side of the nano-oncology field focuses on developing new structures that able to perform efficient target therapy. Therefore, this type of NPs disposes of unique physical and chemical properties for active targeting of the desired cells providing imaging and therapeutic action against the disease [8].

2. Nanoparticles

The term "nanoparticles" is intensively used in the nanomedicine field in order to describe a particle with a size in the range of 1–100 nm. NPs are designed from a wide class of materials, including metals, silicates, metal oxides, polymers, organics, non-oxide ceramics, carbon and biomolecules. For biomedical applications, NPs are presented in different morphological states such as spheres, tubes, cylinders, platelets [9].

NPs have surface modifications that can facilitate the internalization/uptake of therapeutic agents and also their capability to travel through the bloodstream to the target sites. Generally, the structure of NPs is composed of three different layers, including the surface layer (can be functionalized with a wide range of small molecules, surfactants, metal ions and polymers), the shell layer (consists of different chemical material according to the core of the NPs) and the core (represents the central portion of the NP) [10]. Therefore, NPs have exceptional characteristics due to their structure and design and gained an enormous interest in multidisciplinary fields such as drug delivery [11], cancer therapy, tissue engineering, protein detection, multicolor optical coding for biological assays, manipulation of cells and biomolecules [12], imaging, biosensors, hyperthermia, photoablation therapy and gene delivery [13]. They exhibit special physical and chemical properties like a high surface area-to-volume ratio and also a unique quantum size effect superior to their corresponding bulk materials. Moreover, NPs' controllable size and shape play an important role in medical applications [14]. Moreover, there are some nanomaterials that can exhibit intrinsic therapeutic properties such as gold nanoshells, which have the potential to deliver photothermal therapy [15].

Currently, the term "theranostics" starts to gain attention in the medical and research field, and it describes single biocompatible and biodegradable nanoparticle, which can contain both therapeutic and diagnostic compounds (**Figure 2**) [16]. Specifically, theranostic nanoparticles (TNPs) have been designed in order to be applied for multiple imaging approaches including optical imaging, ultrasound (US), magnetic resonance imaging (MRI), computed tomography (CT), single-photon computed tomography (SPECT) and positron emission tomography (PET) [17]. Moreover, TNPs are able to improve the accumulation and delivery of the active compounds at the tumor site, enhancing therapeutic efficacy and reducing the intensity of side effects on healthy tissues [18], and they can be eliminated from the body in a short period of time and degrade into nontoxic bioproducts [19].

2.1 Synthesis of NPs

Synthesis of NPs can be performed using various methods, which are divided into two main classes such as bottom-up (chemical synthesis) and top-down (mechanical attrition) approaches (**Figure 3**) [20]. Bottom-up method is based

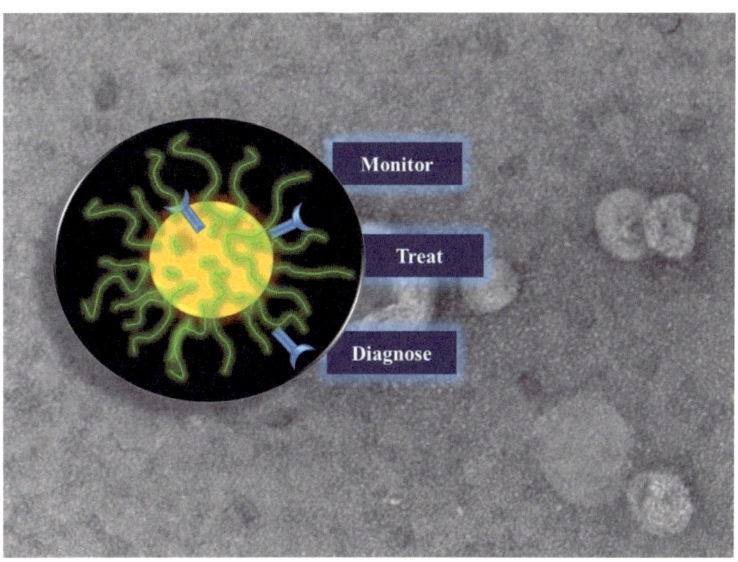

Figure 2.
Theranostic nanoparticles used in the medical field in order to improve the diagnosis and therapeutic approaches.

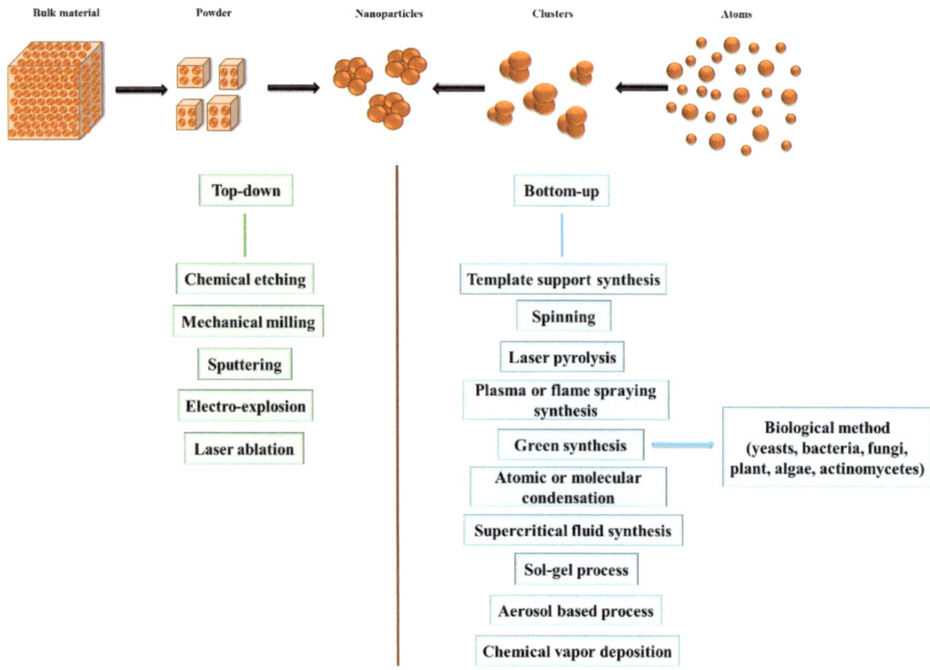

Figure 3.
Common methods used to synthesis NPs via top-down and bottom-up approaches.

on larger nanostructures design beginning from smaller building blocks including atoms and molecules. Meanwhile, the top-down approach refers to larger molecules, which are decomposed into smaller building blocks and then converted into suitable NPs [10]. Traditional chemical and physical methods present some main drawback due to the presence of reducing and stabilizing agents, which carry a risk of toxicity to the environment and also to the cell [21].

Currently, green chemistry has been suggested as a valuable alternative for metal nanoparticles synthesis that employs biological entities including microorganisms and plant extracts [22]. The main role of microorganisms (bacteria and fungi) is involved in the remediation of toxic materials by reducing metal ions [23]. The most often used metal for green synthesis is silver, gold, iron, and copper [24]. Therefore, the size distribution of NPs is strongly depended on the presence of the biocompounds, which are found in the extract. These biocompounds (phenolic compounds, alkaloids, enzymes, terpenoids, proteins, co-enzymes, sugar and others) are mainly involved in reducing the oxidative state of the metal salts from positive to zero oxidative state [25]. Few bacteria have been shown the potential to synthesize silver nanoparticles intracellularly where intracellular components have the ability to act as reducing and stabilizing agents, respectively [26]. Thus, the green synthesis of nanoparticles could be a promising approach to replace many complex physiochemical syntheses due to their advantages such as no need to use toxic chemicals, free from hazardous by-products and also the use of natural capping agents [27].

In their study, Mirtaheri et al. had succeeded in synthesis of mesoporous tungsten oxide using a template-assisted sol-gel method, which relies on the photocatalytic degradation of Rhodamine B [28]. Mesoporous TiO_2-SiO_2 were synthesized by Haghighatzadeh et al. using an ultrasonic impregnation method. In addition, under 800°, they synthesized the anatase crystals with higher photocatalytic efficiency for degradation of methylene blue [29]. Deshmukh et al. synthesized various nanoparticles using plant extracts in order to evaluate their antibacterial and antioxidant activity for targeted applications [30]. Another study on this topic is showed by Baltazar-Encarnacion et al., which described the green synthesis of Ag nanoparticles using an *E. coli* for the production of NPs with antimicrobial properties against bacteria [31]. Green biosynthesis methods are more reliable and safer than chemical synthesis [32].

Structural DNA nanotechnology is a precise method, which is used to control the NPs shape. In particular, the DNA-origami method allows the controlled self-assembly of 2D and 3D nanostructures with nanometer precision [33]. Such nanoparticles can be used to detect short oligonucleotides in a microbead-based assay [34] and can be applied in the biological field, nanoelectronics and nanophotonics [35]. Therefore, these designs provide comprehensive understanding of cellular interactions regarding tumor detection strategies [36, 37].

Specifically, TNPs can be engineered in several ways. For example, TNPs can be obtained by conjugating therapeutic agents (chemotherapy and photosensitizers) to existing imaging NPs (quantum dots, gold nanocages and iron oxide NPs). On the other hand, NPs can encapsulate both imaging and therapeutic agents in biocompatible nanosystems such as ferritin nanocages, polymeric and porous silica NPs. Other unique NPs such as porphycenes, [64Cu] CuS, gold nanoshells or cages have inherent imaging and therapeutic characteristics [19].

2.2 Characterization of NPs

Physicochemical properties of NPs (shape, size, composition, optics) can be analyzed through different techniques.

The morphology of NPs is characterized through microscopic techniques including polarized optical microscopy (POM), transmission electron microscopy (TEM) and scanning electron microscopy (SEM), which are the most relevant techniques in this area. SEM technique provides relevant information regarding the nanoscale level of the NPs [38]. Moreover, TEM provides features about the bulk material used for NPs synthesis at very low to higher magnification [39]. The morphological features of the NPs exhibit a relevant interest since their morphology influences the NP's properties [10].

Structural characterization is based on the study of the composition and nature of bonding materials. The common techniques used to study the bulk properties are X-ray diffraction (XRD), X-ray photoelectron spectroscopy (XPS), infrared spectroscopy (IR), Raman, Brunauer-Emmett-Teller (BET), energy dispersive X-ray (EDX) and Zeta size analyzer. Through XRD technique, the crystalline structure and the phase of the NPs are identified. The most sensitive technique used to characterize NPs is XPS, which determines the exact element ratio and bonding nature of the elements used for NPs synthesis [10].

Optical characterizations are widely used to obtain information about the absorption, reflectance, phosphorescence and luminescence of NPs. This method is based on the Beer-Lambert law and basic light principles. These properties are highlighted through several techniques, including diffuse reflectance spectroscopy (DRS), UV and UV-Vis, which reveal good knowledge about the mechanism of their photochemical processes [10].

2.3 Physicochemical properties of NPs

For cancer research, NPs can be modified respecting the size, shape and surface to improve their ability to reach tumors. Smaller NPs have the ability to accumulate more easily in the leaky blood vessels of tumor sites compared to larger NPs, which can remain at the injection site [40].

Nowadays, ultrasmall nanoparticles (1–3 nm cores) are widely used for medical applications because of their advantages regarding biodistribution, targeting features, adsorption, easy surface modifications and pharmacokinetics [18, 41–43]. Gadolinium ultrasmall nanoparticles achieved theranostic potential without considerable toxicity *in vivo* in the case of brain cancers [44]. Another example is represented by ultrasmall silica nanoparticles functionalized with antibody fragments used to target HER2-overexpressed breast cancer as imaging agents [42].

Metallic nanoparticles can be designed as ultrasmall constructs too. In this regard, it is important to mention D-peptide p53 activator gold nanoparticle conjugates used for cancer target therapy [45], bimetallic nanoparticles for triggered ultrasound cancer therapy [46] and Cu ultrasmall nanoparticles' valuable ability for photothermal cancer therapy [47].

On the other hand, NP shape influences the fluid dynamics and uptake into tumor sites. Non-spherical NPs present excellent optical properties due to surface plasmon resonances and are strongly recommended for cancer phototherapy applications [48–50]. Furthermore, rod-like shape nanoparticles are better accepted and tolerated by the organism [51, 52].

Specifically, spherical NPs started to be more common than non-spherical NPs due to challenges in synthesis approaches and testing [53]. Spherical silver nanoparticles ensure anti-inflammatory potential [54] and promote camptothecin apoptotic activity in cervical cancer [55]. Despite the advantages offered by silver nanoparticles, progress in spherical gold nanoparticles makes possible their use for combined therapies like drug delivery and photothermy [56].

There are other significant factors that contribute to a successful therapy development. Stability and distribution are affected by NPs charge. A positive charge is most effective according to tumor vessels targeting, but a switch to a neutral charge allows NPs to diffuse to the tumor sites [57]. In order to prolong blood circulation of NPs, their surface can also be modified with specific molecules (hydrophilic polymers/surfactants, biodegradable copolymers such as polyethylene glycol, poloxamine, polyethylene oxide and polysorbate 80), which facilitate cellular uptake into tumor tissue [58, 59].

2.4 Classification of NPs

Modern nanosystems can enhance drug diagnosis, delivery and also monitor therapeutic responses to the provided drugs [60]. In order to improve clinical outcomes, researchers tried to synthesize a theranostic platform consisting of multifunctional NPs, which exhibit valuable imaging properties. Therefore, TNPs can be composed of lipids, polymers, metals, carbon and ceramics [61].

Lipid nanoparticles are widely used in medical field due to their biodegradability, biocompatibility, low toxicity and high loading capacity for both hydrophobic and hydrophilic drug molecules [62, 63]. Moreover, they can improve the pharmacodynamics and the pharmacokinetics of therapeutic agents based on controlled release profile [64]. Another important characteristic of lipid NPs is their availability for functionalization with antibodies, peptides, small molecules or aptamers in order to perform target therapy [65–67].

Polymeric NPs are normally organic-based NPs with a diameter lower than 1 μm. They can be called nanospheres or nanocapsules depending on their composition [68–70]. These nanoparticles have the ability to improve both the solubility and the bioavailability of hydrophobic drugs [71] and are intensively used as delivery systems [72, 73].

Metallic NPs are designed from metal precursors, including noble metals (Cu, Ag, and Au). The most researched area in biomedical field is represented by gold

NPs, which possess unique optical and electronic characteristics as well as chemical inertness. Also, their availability for surface functionalization [74–76] makes them very useful for a lot of medical applications such as biosensing [77], bioimaging [78] and photothermal therapy [79]. Silver nanoparticles exhibit unique properties such as thermal conductivity, high electrical conductivity, catalytic activity, chemical stability, antibacterial and improved optical properties [80]. These NPs are suitable for photonic [81], electronic [82], antimicrobial and disinfectant applications [83, 84], biosensors [85], drug delivery, photothermal therapy [26] and cellular imaging [86].

Another class of metallic nanoparticles is represented by semiconductor nanocrystals, which are well known as quantum dots. Many studies report their potential use in biomedical imaging [87], drug and gene delivery [88] and also in diagnosis [89] based on their unique chemical and optical properties.

Magnetic NPs represented by iron oxide NPs possess unique chemical, biological and magnetic characteristics including non-toxicity, chemical stability, biocompatibility, high magnetic susceptibility and high saturation magnetization [90, 91]. The main drawback of iron nanoparticle is that it has a tendency to oxidize [13]. To eliminate this unwanted process, coating with a biocompatible shell, such as a polymer [92], ceramics [93] or metals [13], is needed in order to prevent conglomeration. In addition, iron oxide NPs can be functionalized with proteins, antibodies, enzymes and anticancer drugs [13] and are investigated for different applications including magnetic hyperthermia [94], contrast agents in MRI (magnetic resonance imaging) [95], targeted drug delivery [96], multimodal imaging and gene therapy [61].

In the term of carbon-based NPs, fullerenes and carbon nanotubes exhibit promising biomedical applications. Fullerenes are suitable for multiple functionalization steps according to their particular globular network structure [97]. They are widely used as excellent antioxidants [98], antiviral agents [99, 100], drug and gene delivery systems [101–103] and photosensitizers for photodynamic therapy [104, 105]. On the other hand, elongated design of carbon nanotubes diagnostic imaging strategies [106–110], drug delivery [111–113] and also photothermal therapy [114, 115].

Ceramics NPs are inorganic non-metallic solids, which are synthesized by heating and successive cooling [116]. Therefore, these ceramics NPs are intensively used in the research field as photocatalysis, catalysis, agents for photodegradation of dyes and imaging agents [117].

There are significant challenges in engineering and designing new nanosystems. The "nanoparticle loaded nanoparticle" concept is described as an innovative strategy composed of at least two different nanoparticles. For example, porous nanoparticles made by silica can encapsulate DNA-conjugated small gold nanoparticles in their pores with great applicability in penetrating tumors [118].

Hybrid constructs gained increased interest in obtaining programmed nanoparticles. DNA nanorobots built of a DNA robot and a DNA aptamer that confers molecular recognition of nucleolin are used for target therapy in cancer [119].

3. Cellular internalization and endosomal escape

Once the delivery system comes in the proximity of its target site, the drug must be internalized in order to fulfill its biological effect. While free drugs usually have the ability to pass through cellular membranes and accumulate inside the cell unless they are externalized by efflux pump mechanisms, NPs are internalized differently, mainly through various types of endocytosis [120], as presented in **Figure 4**.

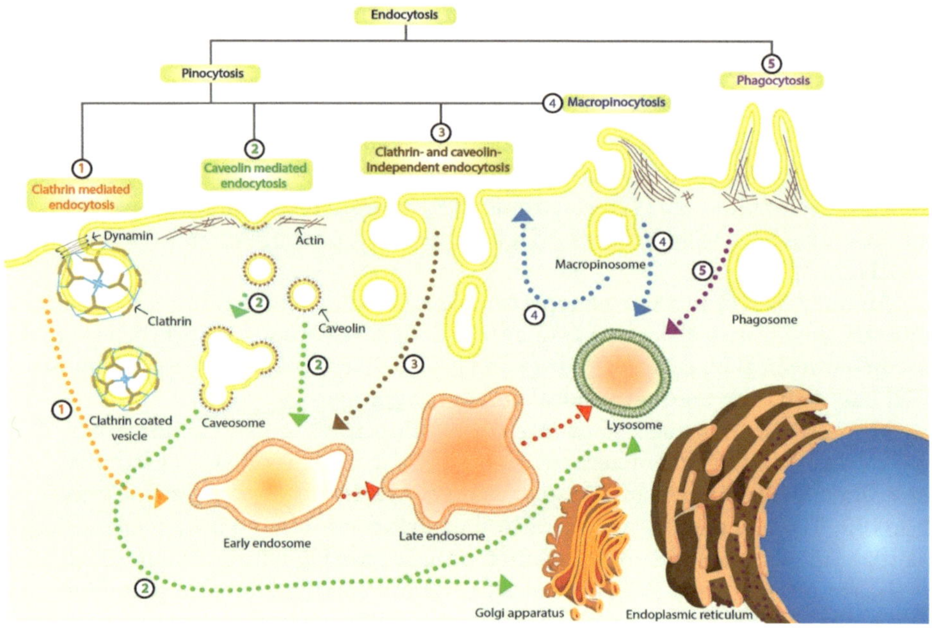

Figure 4.
Cellular internalization through endocytosis.

Phagocytosis is a mechanism by which specialized cells known as phagocytes recognize and engulf large particles (≥0.5 μm) into vesicles called phagosomes [121]. This process involves actin polymerization and the extension of pseudopods, which surround the opsonized target object [122] leading to its internalization (**Figure 4(5)**). Phagosomes fuse with early endosomes, followed by late endosomes and then lysosomes, becoming highly acidic and possessing hydrolytic enzymes leading to the degradation of the engulfed object [122].

Macropinocytosis is a process by which nonselective molecules suspended in extracellular fluid are internalized into the cell, giving rise to endocytic vesicles. Like phagocytosis, it involves cytoskeleton rearrangement beneath the plasma membrane. This leads to a plasma membrane circular ruffle formation that extends and entraps extracellular material, producing a so-called macropinosome [123]. The maturation of these vesicles involves shrinking while concentrating their contents, migration and digestion or recycling of their contents [124]. Depending on the cell line, macropinosomes can fuse with lysosomes or directly to the plasma membrane expelling their content to the extracellular space (**Figure 4(4)**) [124, 125].

Caveolae are small (60–80 nm) plasma membrane invaginations, important in processes such as endocytosis, transcytosis, potocytosis and certain signaling pathways [126]. Caveolin-dependent endocytosis is a triggered, energy-dependent event involved in the uptake of extracellular molecules and membrane components [127]. It is dependent on actin and dynamin, a GTPase, which is present at the neck of caveolae and is responsible for the release of the caveolar vesicle inside the cytoplasm [128]. These vesicles deliver the internalized molecules to caveosomes or to early endosomes (**Figure 4(2)**). Caveosomes bypass lysosomes, thus being an important approach for administering easily degradable therapeutic agents [129].

Clathrin-mediated endocytosis involves the uptake of extracellular molecules through invagination of the plasma membrane. The vesicles are formed when ligands interact with receptors on the plasma membrane, thus recruiting clathrin triskelions and adaptor proteins, which form a multifaceted cage structure [130]

Strategy		**Mechanism**	**Examples**	**Ref.**
Endosomal membrane destabilization	pH dependent	pH buffering (proton sponge effect)	Polyamines (PEI, PEAAc, Mglu-HPG)	[134]
		Pore-formation	Listeriolysin O (LLO) GALA peptide	[135]
	pH independent	Pore formation	Amphotericin B Melittin	[136, 137]
Fusion with endosomal membrane		Flip-flop mechanism	GALA peptide	[138, 139]
		Via viral fusion proteins/ peptides	HA2 fusion peptide/ hemagglutinin	[140]
Photochemical membrane rupture		Light-induced ROS and/ or heat generation	TatU1A-photosensitizer conjugates M-PLL (melanin-poly-L-lysine)	[141, 142]

Table 1.
Approaches for endosomal escape.

that is released inside the cell with the help of dynamin. These vesicles are known as clathrin-coated vesicles and can lose their clathrin coat and fuse with early endosomes (**Figure 4(1)**). They are directed towards degradation in lysosomes or recycled to the plasma membrane [131].

Extracellular cargo can also be internalized via clathrin- and caveolin-independent pathways (**Figure 4(3)**) [132].

Depending on the internalization mechanism, NPs have different fates. They can face lysosomal degradation when internalized through clathrin-mediated endocytosis while skipping this process when taken up through a caveolin-mediated mechanism [133].

Many nanomaterials are degraded in endocytic vesicles leading to new approaches of carrier designs that are able to escape the endosomal or lysosomal degradation. Three main strategies, presented in **Table 1**, are commonly used to bypass this cellular barrier for drug administration. They rely on molecules, which possess the ability to destabilize the endosomal membrane in a pH-dependent or independent way or to fuse with the endosomal membrane, leading to the release of previously internalized cargo. Another approach involves the photochemical membrane rupture via photothermal nanomaterials.

4. Diagnosis through molecular imaging mechanisms

Molecular imaging is a medical discipline related to medical imaging and is representing the evolution of imaging techniques for diagnosis and therapy monitoring. It involves cell biology and molecular biology [143].

Current clinical applications of molecular imaging are CT, SPECT, PET, MRI, US and also hybrid imaging techniques SPECT/CT, PET/CT or PET/MRI. CT, MRI and US provide anatomical information, while PET/CT, and SPECT/CT offer functional and molecular information [144]. All these techniques are based on the accumulation of a contrast agent at the target site [145].

Even if they provide high-resolution images from anatomical [146] to molecular level for further clinical investigations [147–153], there are some disadvantages regarding the use of them. High doses of radiation and exposure can cause DNA damage in some tissues [154, 155]. Also, radiopharmaceutical biodistribution and effectiveness may cause image artifacts and also side effects for the patient [156–159]. Moreover, the patient care quality is not granted in most of the cases [156].

4.1 NPs involved in diagnosis imaging strategies

Diagnostic imaging using NPs refers to the detection of specific disease sites through molecular recognition of tumor cell particularities like the overexpression of several genes and the presence of different cell surface molecules or media excreted compounds/molecules that are involved in various disease processes, microenvironment particularities and also cell development stages [160, 161].

Physical properties of nanoparticle systems are very important for molecular imaging applications. Nanoparticle accepted diameters for this application are between 30 and 150 nm. Usually, the nanoparticle surface is modified using a ligand in order to target specific tumor cell molecules. As more ligands are attached on the nanoparticle surface, there are more chances to bind the target cell. The amount of signaling groups influence the sensitivity of the detection method [145].

Some NPs have innate optical properties like QDs [162] and metallic NPs due to surface plasmon resonance [48, 163–165]. QDs nanoparticles labeled with ^{18}F-Fluoropropionate and functionalized with RGD peptides demonstrate proper optical characteristics for PET imaging of prostate cancer [166].

Gold nanoparticles proved long circulation time and useful optical properties like high spatial resolution and high sensitivity for CT imaging. By functionalization with chitosan polymers, they were used for colorectal adenocarcinoma imaging [167]. Also, they were conjugated with antibodies for lymph nodes and metastases imaging in squamous cell carcinoma, head and neck cancer [168]. Moreover, gold nanoparticles radiolabeled with ^{111}In and ^{125}I can be used in SPECT imaging of epidermoid carcinoma [169].

Iron oxide nanoparticles are widely used in MRI imaging because they can improve and enhance the contrast [170]. In glioblastoma, iron oxide nanoparticles functionalized with peptides and polymers accumulate within tumor microenvironment by forming self-assembly structures [171].

Furthermore, polymeric materials such as mesoporous silica nanoparticles carry tumor targeting properties and are proposed for PET imaging in breast cancer. Besides this, they are able to perform drug delivery applications [172].

Regarding US imaging, perfluorocarbon nanoparticles can be used for a real-time and non-invasive analysis of thyroid carcinoma [173].

Considering the other nanoparticle formulations (nanoliposomes, micelles, polymersomes, dendrimers and aptamers), these ones need to be functionalized with specific contrast agents and fluorophores. The advantages to implement NPs such as molecular imaging tools are biocompatibility and biodegradability [174], encapsulation properties [175], water solubility in some cases [176] and targeting ligands accessibility [177].

Fluorophores are widely used in diagnosis applications and imaging of cellular processes. One drawback of conventional fluorophores is represented by the loss of fluorescence after a long exposure to light, known as photobleaching.

Various processes are known to induce the molecular relaxation without the emission of light, which depends on different chemical or physical factors like temperature, pressure, the presence of organic molecules or polymers and ionic strength, resulting in a decrease in the fluorescence intensity, referred to as quenching [178]. Quantifying this decrease in fluorescence emission can give information about the concentration of a specific compound in the proximity of the nano-objects. Lately, numerous diagnostic techniques based on this phenomenon have been introduced [179, 180].

On the other hand, another luminogen system based on a process called aggregation-induced emission (AIE), developed by Ben Zhong Tang's group in 2001 [181], gathered increased interest for imaging and theranostic applications. Most luminescent systems have a lower efficiency in an aggregated state, thus limiting the concentration that can be used for imaging purposes and at the same time the achievable intensity of the emitted light. However, in the case of AIEgens, aggregation works constructively becoming highly luminescent in concentrated solutions or in an aggregated state. The utilization of AIEgens in theranostics has lately become a reliable approach, because of several advantages that include good biocompatibility, excellent optical properties and simple preparation and conjugation [182]. One example implies the conjugation of an AIEgen (TPS) with a short peptide (DEVD) that is susceptible to caspase-3 cleavage and that is bound to a prodrug that induces apoptosis [183].

5. Targeted therapy

Targeted therapy is a form of treatment, which implies the ability of a drug to accumulate at a target site in the body and thus decrease the side effects in healthy cells and tissues. Nanocarriers are often used to improve the bioavailability of the active compounds at the target site and allow the use of significantly reduced concentrations, therefore limiting the exposure of normal cells to the toxic effects of the drugs [184].

The most common strategies for drug delivery include local drug delivery, passive targeting, physical targeting, magnetic targeting and active targeting [185].

Local drug delivery is a promising strategy for the treatment of metabolic disorders (diabetes and obesity) [186], periodontitis [187] and bone disorders [188] due to its potential to keep drug availability in the target site for a prolonged period of time.

Passive targeting is based on enhanced permeability and retention effect (EPR effect) present in many tissues [189, 190]. Macromolecules and NPs from the bloodstream accumulate preferentially in tumors and inflamed sites, where the permeability of the vasculature is often enhanced. Moreover, the lymphatic drainage system is damaged in tumors, leading to a prolonged retention of the macromolecules and NPs in the tumor interstitium [191].

Physical targeting depends on the optical, thermal and electrical properties of the carriers [192], which can disintegrate at lower pH values or higher temperature and release the free drug. The tumor microenvironment is more acidic compared to the normal surrounding tissues, due to the accumulation of lactate, and therefore provides an opportunity for the use of pH-sensitive nanocarriers in cancer therapy [193].

Another approach for drug targeting refers to the accumulation of superparamagnetic carriers in target sites under the action of external magnetic field. Thus, a larger dose of the drug can be released at the tumor site for an increased period of time and side effects of chemotherapy can be diminished [194]. Once systemically administered, besides the type and intensity of the magnetic field and size of the NPs, many biological factors influence the infiltration of the superparamagnetic carriers to the target site, including the effect of Brownian motion, blood viscosity, interaction of the particles with the red blood cells and blood matrix [195].

While in the case of passive targeting the physicochemical properties of the nanocarrier system play the major role, active targeting relies on the interaction between the surface of the carrier and antigens expressed on target cells. NPs are functionalized

Class	Ligand	Targeted biomarker	Disease (clinical trials = *)	Ref.
Antibodies	Trastuzumab, cetuximab, Anti-CD20 mAbs (Rituximab)	HER2 receptor, EGFR, CD20	Breast cancer*, esophageal carcinoma*, pancreatic adenocarcinoma*, head and neck cancer*, non-Hodgkin's lymphoma*, rheumatoid arthritis*	[197–203]
Peptides	Transferrin	Transferrin receptor	Cancer	[204, 205]
Small molecules	Folic acid	Folate receptor	Rheumatoid arthritis*, ovarian cancer, lung cancer*	[206, 207]
Aptamers	A10RNA, AS1411, Anti-MUC1	Extracellular domain of the PSMA, nucleolin, MUC1	Prostate cancer, breast cancer	[208–210]

**Refers to clinical studies.*

Table 2.
Commonly used molecules for active targeting.

by adsorption or chemical conjugation with a large variety of ligand types such as peptides, small molecules, proteins, and aptamers, which present a high specificity for epitopes or receptors that are uniquely expressed or overexpressed on the target sites [196]. Examples of commonly used ligands and their targets are presented in **Table 2**.

6. Theranostic NPs recently developed

Theranostics refers to the use of the nanoparticle for molecular imaging and therapy. Considering the biological barriers, the biocompatibility, easy surface modifications, controlled pharmacokinetics and biodistribution and accommodation in various microenvironment conditions are still necessary to be accomplished [211]. Polymers are widely used for NP formulations because of biocompatibility and biodegradability properties *in vivo* [212–215]. Besides coating the nanoparticle surface with polymers, the fluorophores and other contrast probes are widely used to achieve high-sensitivity molecular imaging.

There are three main theranostic directions that involve the use of nanoparticles. The first strategy refers to treatment effect evaluation through molecular imaging with NPs as contrast agents. The aim of the second one is to assess a nanoparticle therapeutic strategy with molecular imaging probes. The third one describes nanoparticles as target therapy agents and molecular imaging tools at the same time. In this regard, for the first two procedures, the NP system is either the evaluator or the evaluated component, and for the last strategy, these roles are overlapping. Each one of these roles makes possible the development of future therapies (**Figure 5**).

The nanoparticles' evaluator role (**Figure 5(1)**) can be emphasized in the next study. Zhang et al. developed Annexin A5-conjugated polymeric micelles with dual role: detection of apoptosis via SPECT and optical imaging and also therapy outcomes investigation. In this study, the apoptosis was induced by drugs like cyclophosphamide, etoposide, poly (L-glutamic acid)-paclitaxel and cetuximab (IMC-C225) anti-EGFR antibody. The NPs were used to observe the apoptosis-induced processes in lymphoma and breast cancer *in vivo*. Therefore, SPECT and fluorescence molecular tomography allowed cellular death visualization in tumors [216].

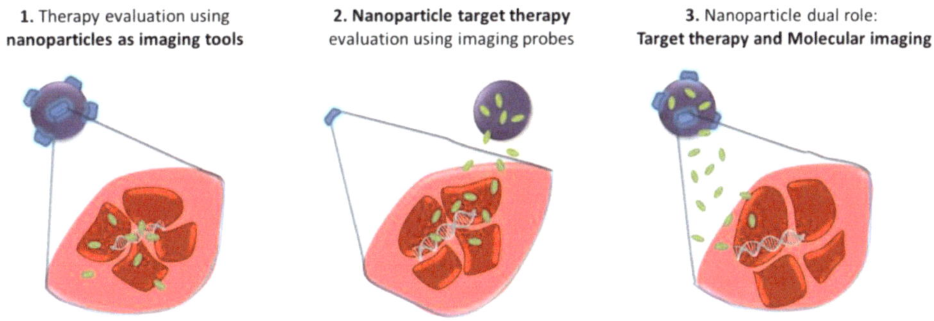

Figure 5.
Theranostic NPs action strategies.

NP effect evaluation (**Figure 5(2)**) can be performed based on probes that are currently used in clinical molecular imaging. For example, 2-deoxy-2-[F-18]fluoro-D-glucose (^{18}F-FDG) probe is used for metabolic activity measurements via PET/CT imaging. This radiolabeled probe can act as prognostic biomarker for nanoparticle-assisted photothermal therapy monitoring in neuroendocrine lung cancer *in vivo* [217].

Another strategy is to ensure both imaging and therapy at the same time (**Figure 5(3)**). In this situation, the nanosystem can be composed of two different components bonded together in order to perform a theranostic action.

The easiest way is to make use of the optical properties developed by some materials at nanoscale. Therefore, metallic nanoparticles can scatter and absorb the light in the NIR wavelength domain and are promising tools for cancer photothermal therapy [218].

In a different way, nanoparticles can be associated with molecular imaging techniques in order to enhance their efficiency. For example, doxorubicin-loaded polymeric micelles and perfluoropentane stabilized by the same block copolymer can perform US imaging and target therapy for breast and ovarian cancer [219, 220].

Some designs suggest the use of two different nanoparticles, which by conjugation with targeting ligands and drug molecules provide tumor visualization and target therapy. For example, quantum dot-mucin 1 aptamer-doxorubicin conjugates were used for ovarian cancer targeting and proved suitable optical properties for imaging and controlled release of the drug [221].

In addition to the molecular imaging techniques previously described, some nanoparticles can be used for photodynamic and photothermal therapy in order to perform targeting therapy.

Photodynamic therapy (PDT) implies the use of photosensitizer agents that under laser irradiation exert cytotoxic activity by generating reactive oxygen species [222, 223]. This therapy is very appreciated regarding multidrug resistance cancers and is suppose that it can replace the conventional chemotherapy [224]. PDT-specific nanoparticles are used as photosensitizer carriers [225, 226]. Moreover, these nanocarriers can be functionalized with targeting ligands for better tumor selectivity and also with drug molecules for therapeutic effectiveness [227–229]. Gold nanoparticles loaded with a fluorescent drug Pc4 targeting PSMA-1 membrane antigen in prostate cancer are promising tools for surgical guidance and further therapeutic intervention [228]. EGFR-targeted liposomal nanohybrid cerasomes are proposed for PDT and immunotherapy in colorectal cancer due to their sensitive detection properties and anti-tumor efficacy [229].

By a theranostic point of view, photothermal therapy (PTT), also known as hyperthermia or thermal ablation therapy, acts as a diagnosis and a treatment

Stage	Nanoparticle type	Therapeutic agent	Diagnostic agent	Pathology	Target	Ref.
Pre-clinical	Liposomes (100–200 nm)	Paclitaxel	pH-sensitive poly(ethylene oxide) (PEO)-modified poly(beta-amino ester) (PbAE) nanoparticles	Ovarian adenocarcinoma	EPR	[235]
	Silica (100–200 nm)	Paclitaxel and camptothecin	Superparamagnetic iron oxide nanocrystals	Pancreatic cancer	Folic acid	[236]
	Iron oxide (10–25 nm)	Anti-EGFRIgG	Iron oxide nanoparticles	Glioblastoma	EGFR	[237]
	Gold nanorod (10 x 40 nm)	Heat	Thermal/CT	Breast cancer	EPR	[238]
	Quantum dots (30–50 nm)	Paclitaxel, doxorubicin, 5-fluorouracil	Quantum dots	Many cancers	CD44, folic acid	[239]
Clinical trials	Silica (6–7 nm)	cRGDY	Ultrasmall inorganic hybrid nanoparticles	Melanoma and malignant brain tumors	$\alpha_v\beta_3$ integrin	[240]
	Cyclodextrin (70 nm)	RNAi	Transferrin	Solid tumors	Transferrin receptor	[241]
	Silica-gold nanoshell	Photothermal ablation	Nanoshell (MR and optical)	Head/neck cancer, primary and/or metastatic lung tumors	EPR	[242]
	Gold (27 nm)	Tumor necrosis factor alpha	Gold nanoparticles	Solid tumors	EPR (passive mechanism) rhTNF (active mechanism)	[243]
	Iron oxide	Endorem (superparamagnetic particles of iron oxide)	Iron oxide	Healthy volunteers	none	[244]

Abbreviations: EPR, enhanced permeability and retention effect; EGFR, epidermal growth factor receptor; cRGDY, peptide cyclo-(Arg-Gly-Asp-Tyr); rhTNF, recombinant human tumor necrosis factor alpha; RNAi, ribonucleic acid interference; MR, magnetic resolution.

Table 3.
Nanoparticles used in clinical (according to clinicaltrials.gov) and pre-clinical work.

strategy. It uses electromagnetic radiation in infrared (IR) region and provides high specificity analysis and minimal invasiveness [230]. The nanocarriers used for PTT need to have the capacity to target the tumor site after heat generation under laser irradiation [231]. For this purpose, various drug molecules and targeting ligands are encapsulated into nanoparticles. Gold nanoshells targeting HER2 positive breast cancer proved optical contrast and high tissue penetration under NIR irradiation [218]. Polymer nanoparticles functionalized with IR820 and doxorubicin were used in ovarian cancer and showed prolonged circulation time and drug accumulation at the target site [232]. It is important to mention that the generated temperature is usually between 42 and 45°C and sometimes higher depending on tumor tissue [233, 234].

6.1 Theranostic nanoparticles used in the clinic

There are various types of theranostic NPs that can be designed and used for cancer diagnosis and therapy. Their applicability is highlighted by liposomes, which are intensively used in clinical trials due to their specific features. In **Table 3**, several theranostic nanoparticles used in clinical (clinical trials) and pre-clinical work for cancer diagnosis and therapy are shown.

Theranostics has the potential to predict and evaluate therapy response, offering advantageous opportunities to modify the ongoing treatments and to develop new ones even in a personalized manner [245]. Nanoparticles have gained a lot of confidence in becoming important tools for a lot of medical applications due to their properties [17, 19].

The newest designs focus on hybrid nanostructures for better sensitivity and accuracy. These nanohybrids are currently studied and they proved effectiveness in cancer targeting by combining different imaging techniques with drug delivery strategies [246–248].

Acknowledgements

This research was funded by the research grants "Clinical and economic impact of personalized targeted anti-microRNA therapies in reconverting lung cancer chemoresistance"-CANTEMIR, POC-P-37-796/2016, "Innovative advanced approaches for predictive regenerative medicine"—REGMED, no. 65PCCDI/2018, PN-III-P1-1.2-PCCDI-2017-0782, "Increasing the performance of scientific research and technology transfer in translational medicine through the formation of a new generation of young researchers"—ECHITAS, no. 29PFE/18.10.2018, PNCDI III 2015-2020.

Conflict of interest

The authors declare no conflict of interest.

Author details

Anca Onaciu[1], Ancuta Jurj[2], Cristian Moldovan[1,2] and Ioana Berindan-Neagoe[1,2,3]*

1 MEDFUTURE-Research Center for Advanced Medicine, "Iuliu Hațieganu" University of Medicine and Pharmacy, Cluj-Napoca, Romania

2 Research Center for Functional Genomics, Biomedicine and Translational Medicine, "Iuliu Hațieganu" University of Medicine and Pharmacy, Cluj-Napoca, Romania

3 Department of Functional Genomics and Experimental Pathology, The Oncology Institute "Prof. Dr. Ion Chiricuta", Cluj-Napoca, Romania

*Address all correspondence to: ioananeagoe29@gmail.com

References

[1] Singh NA. Nanotechnology innovations, industrial applications and patents. Environmental Chemistry Letters. 2017;**15**:185-191. DOI: 10.1007/s10311-017-0612-8

[2] Nikolaou M, Pavlopoulou A, Georgakilas AG, Kyrodimos E. The challenge of drug resistance in cancer treatment: A current overview. Clinical & Experimental Metastasis. 2018;**35**:309-318. DOI: 10.1007/s10585-018-9903-0

[3] Ventola CL. Progress in nanomedicine: Approved and investigational nanodrugs. P T. 2017;**42**:742-755

[4] Riggio C, Pagni E, Raffa V, Cuschieri A. Nano-oncology: Clinical application for cancer therapy and future perspectives. Journal of Nanomaterials. 2011;**2011**:1-10. DOI: 10.1155/2011/164506

[5] Luo X, Morrin A, Killard AJ, Smyth MR. Application of nanoparticles in electrochemical sensors and biosensors. Electroanalysis. 2006;**18**:319-326. DOI: 10.1002/elan.200503415

[6] Kim J, Lee N, Hyeon T. Recent development of nanoparticles for molecular imaging. Philosophical Transactions of the Royal Society A: Mathematical, Physical and Engineering Sciences. 2017;**375**:20170022. DOI: 10.1098/rsta.2017.0022

[7] Hasan A, Morshed M, Memic A, HassanS,WebsterT,MareiH.Nanoparticles in tissue engineering: Applications, challenges and prospects. International Journal of Nanomedicine. 2018;**13**:5637-5655. DOI: 10.2147/IJN.S153758

[8] Lymperopoulos G, Lymperopoulos P, Alikari V, Dafogianni C, Zyga S, Margari N. Application of Theranostics in Oncology. Cham: Springer; 2017. pp. 119-128

[9] Mudshinge SR, Deore AB, Patil S, Bhalgat CM. Nanoparticles: Emerging carriers for drug delivery. Saudi Pharmaceutical Journal. 2011;**19**:129-141. DOI: 10.1016/j.jsps.2011.04.001

[10] Khan I, Saeed K, Khan I. Nanoparticles: Properties, applications and toxicities. Arabian Journal of Chemistry. 2017;**10**(4). DOI: 10.1016/J.ARABJC.2017.05.011

[11] Bahrami B, Hojjat-Farsangi M, Mohammadi H, Anvari E, Ghalamfarsa G, Yousefi M, et al. Nanoparticles and targeted drug delivery in cancer therapy. Immunology Letters. 2017;**190**:64-83. DOI: 10.1016/j.imlet.2017.07.015

[12] Salata O. Applications of nanoparticles in biology and medicine. Journal of Nanobiotechnology. 2004;**2**:3. DOI: 10.1186/1477-3155-2-3

[13] McNamara K, Tofail SAM. Nanoparticlesinbiomedicalapplications. Advances in Physics: X. 2017;**2**:54-88. DOI: 10.1080/23746149.2016.1254570

[14] Mauricio MD, Guerra-Ojeda S, Marchio P, Valles SL, Aldasoro M, Escribano-Lopez I, et al. Nanoparticles in medicine: A focus on vascular oxidative stress. Oxidative Medicine and Cellular Longevity. 2018;**2018**:1-20. DOI: 10.1155/2018/6231482

[15] Roy Chowdhury M, Schumann C, Bhakta-Guha D, Guha G. Cancer nanotheranostics: Strategies, promises and impediments. Biomedicine & Pharmacotherapy. 2016;**84**:291-304. DOI: 10.1016/j.biopha.2016.09.035

[16] Janib SM, Moses AS, MacKay JA. Imaging and drug delivery using

theranostic nanoparticles. Advanced Drug Delivery Reviews. 2010;**62**:1052-1063. DOI: 10.1016/j.addr.2010.08.004

[17] Zavaleta C, Ho D, Chung EJ. Theranostic nanoparticles for tracking and monitoring disease state. SLAS Technology (Translating Life Sciences Innovation). 2018;**23**:281-293. DOI: 10.1177/2472630317738699

[18] Jurj A, Braicu C, Pop L-A, Tomuleasa C, Gherman C, Berindan-Neagoe I. The new era of nanotechnology, an alternative to change cancer treatment. Drug Design, Development and Therapy. 2017;**11**: 2871-2890. DOI: 10.2147/DDDT.S142337

[19] Chen F, Ehlerding EB, Cai W. Theranostic nanoparticles. Journal of Nuclear Medicine. 2014;**55**:1919-1922. DOI: 10.2967/jnumed.114.146019

[20] Wang Y, Xia Y. Bottom-up and top-down approaches to the synthesis of monodispersed spherical colloids of low melting-point metals. Nano Letters. 2004;**4**(10):2047-2050. DOI: 10.1021/NL048689J

[21] Iravani S, Korbekandi H, Mirmohammadi SV, Zolfaghari B. Synthesis of silver nanoparticles: Chemical, physical and biological methods. Research in Pharmaceutical Sciences. 2014;**9**:385-406

[22] Saiqa Ikram SA. Silver nanoparticles: One pot green synthesis using *Terminalia arjuna* extract for biological application. Journal of Nanomedicine & Nanotechnology. 2015;**6**:4. DOI: 10.4172/2157-7439.1000309

[23] Arokiyaraj S, Vincent S, Saravanan M, Lee Y, Oh YK, Kim KH. Green synthesis of silver nanoparticles using *Rheum palmatum* root extract and their antibacterial activity against *Staphylococcus aureus* and *Pseudomonas aeruginosa*. Artificial Cells, Nanomedicine, and Biotechnology. 2017;**45**:372-379. DOI: 10.3109/21691401.2016.1160403

[24] Han C, Pelaez M, Nadagouda MN, Obare SO, Falaras P, Dunlop PSM, et al. Chapter 5. The green synthesis and Environmental Applications of nanomaterials. In: Sustainable Preparation of Metal Nanoparticles: Methods and Applications. London: Royal Society of Chemistry; 2012. pp. 106-143. DOI: 10.1039/9781849735469-00106

[25] Roy A, Bulut O, Some S, Mandal AK, Yilmaz MD. Green synthesis of silver nanoparticles: Biomolecule-nanoparticle organizations targeting antimicrobial activity. RSC Advances. 2019;**9**:2673-2702. DOI: 10.1039/C8RA08982E

[26] Patra S, Mukherjee S, Barui AK, Ganguly A, Sreedhar B, Patra CR. Green synthesis, characterization of gold and silver nanoparticles and their potential application for cancer therapeutics. Materials Science and Engineering: C. 2015;**53**:298-309. DOI: 10.1016/j.msec.2015.04.048

[27] Lee S, Jun B-H. Silver nanoparticles: Synthesis and application for nanomedicine. International Journal of Molecular Sciences. 2019;**20**:865. DOI: 10.3390/ijms20040865

[28] Mirtaheri B, Shokouhimehr M, Beitollahi A. Synthesis of mesoporous tungsten oxide by template-assisted sol-gel method and its photocatalytic degradation activity. Journal of Sol-Gel Science and Technology. 2017;**82**:148-156. DOI: 10.1007/s10971-016-4289-4

[29] Haghighatzadeh A, Mazinani B, Shokouhimehr M, Samiee L. Preparation of mesoporous TiO_2-SiO_2 by ultrasonic impregnation method and effect of its calcination temperature on photocatalytic activity. Desalination and Water Treatment. 2017;**92**:145. DOI: 10.5004/dwt.2017.21481

[30] Deshmukh AR, Gupta A, Kim BS. Ultrasound assisted green synthesis of silver and iron oxide nanoparticles using fenugreek seed extract and their enhanced antibacterial and antioxidant activities. BioMed Research International. 2019;**2019**:1-14. DOI: 10.1155/2019/1714358

[31] Baltazar-Encarnación E, Escárcega-González CE, Vasto-Anzaldo XG, Cantú-Cárdenas ME, Morones-Ramírez JR. Silver nanoparticles synthesized through green methods using *Escherichia coli* top 10 (Ec-Ts) growth culture medium exhibit antimicrobial properties against nongrowing bacterial strains. Journal of Nanomaterials. 2019;**2019**:1-8. DOI: 10.1155/2019/4637325

[32] Yu C, Tang J, Liu X, Ren X, Zhen M, Wang L. Green biosynthesis of silver nanoparticles using *Eriobotrya japonica* (Thunb.) leaf extract for reductive catalysis. Materials. 2019;**12**:189. DOI: 10.3390/ma12010189

[33] Bastings MMC, Anastassacos FM, Ponnuswamy N, Leifer FG, Cuneo G, Lin C, et al. Modulation of the cellular uptake of DNA origami through control over mass and shape. Nano Letters. 2018;**18**:3557-3564. DOI: 10.1021/acs.nanolett.8b00660

[34] Choi Y, Schmidt C, Tinnefeld P, Bald I, Rödiger S. A new reporter design based on DNA origami nanostructures for quantification of short oligonucleotides using microbeads. Scientific Reports. 2019;**9**:4769. DOI: 10.1038/s41598-019-41136-x

[35] Kasyanenko N, Varshavskii M, Ikonnikov E, Tolstyko E, Belykh R, Sokolov P, et al. DNA modified with metal nanoparticles: Preparation and characterization of ordered metal-DNA nanostructures in a solution and on a substrate. Journal of Nanomaterials. 2016;**2016**:1-12. DOI: 10.1155/2016/3237250

[36] Arora AA, de Silva C. Beyond the smiley face: Applications of structural DNA nanotechnology. Nano Reviews & Experiments. 2018;**9**:1430976. DOI: 10.1080/20022727.2018.1430976

[37] Perrault SD, Shih WM. Virus-inspired membrane encapsulation of DNA nanostructures to achieve in vivo stability. ACS Nano. 2014;**8**:5132-5140. DOI: 10.1021/nn5011914

[38] Saeed K, Khan I. Preparation and properties of single-walled carbon nanotubes/poly(butylene terephthalate) nanocomposites. Iranian Polymer Journal. 2014;**23**:53-58. DOI: 10.1007/s13726-013-0199-2

[39] Khlebtsov N, Dykman L. Biodistribution and toxicity of engineered gold nanoparticles: A review of in vitro and in vivo studies. Chemical Society Reviews. 2011;**40**:1647-1671. DOI: 10.1039/c0cs00018c

[40] Bregoli L, Movia D, Gavigan-Imedio JD, Lysaght J, Reynolds J, Prina-Mello A. Nanomedicine applied to translational oncology: A future perspective on cancer treatment. Nanomedicine: Nanotechnology, Biology and Medicine. 2016;**12**:81-103. DOI: 10.1016/j.nano.2015.08.006

[41] Park W, Heo Y-J, Han DK. New opportunities for nanoparticles in cancer immunotherapy. Biomaterials Research. 2018;**22**:24. DOI: 10.1186/s40824-018-0133-y

[42] Chen F, Ma K, Madajewski B, Zhuang L, Zhang L, Rickert K, et al. Ultrasmall targeted nanoparticles with engineered antibody fragments for imaging detection of HER2-overexpressing breast cancer. Nature Communications. 2018;**9**:4141. DOI: 10.1038/s41467-018-06271-5

[43] Zarschler K, Rocks L, Licciardello N, Boselli L, Polo E,

Garcia KP, et al. Ultrasmall inorganic nanoparticles: State-of-the-art and perspectives for biomedical applications. Nanomedicine: Nanotechnology, Biology and Medicine. 2016;**12**:1663-1701. DOI: 10.1016/j.nano.2016.02.019

[44] Verry C, Sancey L, Dufort S, Le Duc G, Mendoza C, Lux F, et al. Treatment of multiple brain metastases using gadolinium nanoparticles and radiotherapy: NANO-RAD, a phase I study protocol. BMJ Open. 2019;**9**:e023591. DOI: 10.1136/bmjopen-2018-023591

[45] Bian Z, Yan J, Wang S, Li Y, Guo Y, Ma B, et al. Awakening p53 in vivo by D-peptides-functionalized ultra-small nanoparticles: Overcoming biological barriers to D-peptide drug delivery. Theranostics. 2018;**8**:5320-5335. DOI: 10.7150/thno.27165

[46] Gong F, Cheng L, Yang N, Betzer O, Feng L, Zhou Q, et al. Ultrasmall oxygen-deficient bimetallic oxide MnWO X nanoparticles for depletion of endogenous GSH and enhanced sonodynamic cancer therapy. Advanced Materials. 2019;**31**:1900730. DOI: 10.1002/adma.201900730

[47] Wan X, Liu M, Ma M, Chen D, Wu N, Li L, et al. The ultrasmall biocompatible CuS@BSA nanoparticle and its photothermal effects. Frontiers in Pharmacology. 2019;**10**:141. DOI: 10.3389/fphar.2019.00141

[48] Onaciu A, Braicu C, Zimta A-A, Moldovan A, Stiufiuc R, Buse M, et al. Gold nanorods: From anisotropy to opportunity. An evolution update. Nanomedicine. 2019;**14**(9):1203-1226. DOI: 10.2217/nnm-2018-0409

[49] Brzobohatý O, Šiler M, Chvátal L, Karásek V, Zemánek P. Optical trapping of non-spherical plasmonic nanoparticles. In: Andrews DL, Galvez EJ, Glückstad J, editors. Proceedings of SPIE - The International Society for Optical Engineering. Bellingham, Washington USA; 2014. p. 899909. DOI: 10.1117/12.2041199

[50] Eremin YA, Wriedt T, Hergert W. Analysis of the scattering properties of 3D non-spherical plasmonic nanoparticles accounting for non-local effects. Journal of Modern Optics. 2018;**65**:1778-1786. DOI: 10.1080/09500340.2018.1459911

[51] Kolhar P, Anselmo AC, Gupta V, Pant K, Prabhakarpandian B, Ruoslahti E, et al. Using shape effects to target antibody-coated nanoparticles to lung and brain endothelium. Proceedings of the National Academy of Sciences. 2013;**110**:10753-10758. DOI: 10.1073/pnas.1308345110

[52] Zhao Y, Wang Y, Ran F, Cui Y, Liu C, Zhao Q, et al. A comparison between sphere and rod nanoparticles regarding their in vivo biological behavior and pharmacokinetics. Scientific Reports. 2017;**7**:4131. DOI: 10.1038/s41598-017-03834-2

[53] Truong NP, Whittaker MR, Mak CW, Davis TP. The importance of nanoparticle shape in cancer drug delivery. Expert Opinion on Drug Delivery. 2015;**12**:129-142. DOI: 10.1517/17425247.2014.950564

[54] Singh P, Ahn S, Kang J-P, Veronika S, Huo Y, Singh H, et al. In vitro anti-inflammatory activity of spherical silver nanoparticles and monodisperse hexagonal gold nanoparticles by fruit extract of *Prunus serrulata*: A green synthetic approach. Artificial Cells, Nanomedicine, and Biotechnology. 2017;**46**(8):2022-2032. DOI: 10.1080/21691401.2017.1408117

[55] Yuan Y-G, Zhang S, Hwang J-Y, Kong I-K. Silver nanoparticles potentiates cytotoxicity and apoptotic potential of camptothecin in human cervical cancer cells. Oxidative Medicine and Cellular Longevity. 2018;**2018**:1-21. DOI: 10.1155/2018/6121328

[56] Pedrosa P, Mendes R, Cabral R, Martins LMDRS, Baptista PV, Fernandes AR. Combination of chemotherapy and Au-nanoparticle photothermy in the visible light to tackle doxorubicin resistance in cancer cells. Scientific Reports. 2018;**8**:11429. DOI: 10.1038/s41598-018-29870-0

[57] Stylianopoulos T, Poh M-Z, Insin N, Bawendi MG, Fukumura D, Munn LL, et al. Diffusion of particles in the extracellular matrix: The effect of repulsive electrostatic interactions. Biophysical Journal. 2010;**99**:1342-1349. DOI: 10.1016/j.bpj.2010.06.016

[58] Tran S, DeGiovanni P-J, Piel B, Rai P. Cancer nanomedicine: A review of recent success in drug delivery. Clinical and Translational Medicine. 2017;**6**:44. DOI: 10.1186/s40169-017-0175-0

[59] Guerrini L, Alvarez-Puebla R, Pazos-Perez N, Guerrini L, Alvarez-Puebla RA, Pazos-Perez N. Surface modifications of nanoparticles for stability in biological fluids. Materials. 2018;**11**:1154. DOI: 10.3390/ma11071154

[60] Chen J, Wang D, Xi J, Au L, Siekkinen A, Warsen A, et al. Immuno gold nanocages with tailored optical properties for targeted photothermal destruction of cancer cells. Nano Letters. 2007;7:1318-1322. DOI: 10.1021/nl070345g

[61] Vats S, Singh M, Siraj S, Singh H, Tandon S. Role of nanotechnology in theranostics and personalized medicines. Journal of Health Research and Reviews. 2017;**4**:1. DOI: 10.4103/2394-2010.199328

[62] Deshpande PP, Biswas S, Torchilin VP. Current trends in the use of liposomes for tumor targeting. Nanomedicine. 2013;**8**:1509-1528. DOI: 10.2217/nnm.13.118

[63] Voinea M, Simionescu M. Designing of ?Intelligent? Liposomes for efficient delivery of drugs. Journal of Cellular and Molecular Medicine. 2002;**6**:465-474. DOI: 10.1111/j.1582-4934.2002.tb00450.x

[64] Strebhardt K, Ullrich A. Paul Ehrlich's magic bullet concept: 100 years of progress. Nature Reviews Cancer. 2008;**8**:473-480. DOI: 10.1038/nrc2394

[65] Huwyler J, Drewe J, Krähenbuhl S. Tumor targeting using liposomal antineoplastic drugs. International Journal of Nanomedicine. 2008;**3**:21-29

[66] Riaz M, Riaz M, Zhang X, Lin C, Wong K, Chen X, et al. Surface functionalization and targeting strategies of liposomes in solid tumor therapy: A review. International Journal of Molecular Sciences. 2018;**19**:195. DOI: 10.3390/ijms19010195

[67] Puri A, Loomis K, Smith B, Lee J-H, Yavlovich A, Heldman E, et al. Lipid-based nanoparticles as pharmaceutical drug carriers: From concepts to clinic. Critical Reviews in Therapeutic Drug Carrier Systems. 2009;**26**:523-580

[68] Guterres SS, Alves MP, Pohlmann AR. Polymeric nanoparticles, nanospheres and nanocapsules, for cutaneous applications. Drug Target Insights. 2007;**2**:147-157

[69] Mansha M, Khan I, Ullah N, Qurashi A. Synthesis, characterization and visible-light-driven photoelectrochemical hydrogen evolution reaction of carbazole-containing conjugated polymers. International Journal of Hydrogen Energy. 2017;**42**:10952-10961. DOI: 10.1016/J.IJHYDENE.2017.02.053

[70] Hickey JW, Santos JL, Williford J-M, Mao H-Q. Control of polymeric nanoparticle size to improve therapeutic delivery. Journal of Controlled Release. 2015;**219**:536-547. DOI: 10.1016/j.jconrel.2015.10.006

[71] Palmerston Mendes L, Pan J, Torchilin V. Dendrimers as nanocarriers for nucleic acid and drug delivery in cancer therapy. Molecules. 2017;**22**:1401. DOI: 10.3390/molecules22091401

[72] Chaniotakis N, Thermos K, Kalomiraki M. Dendrimers as tunable vectors of drug delivery systems and biomedical and ocular applications. International Journal of Nanomedicine. 2015;**11**:1. DOI: 10.2147/IJN.S93069

[73] Yang J, Zhang Q, Chang H, Cheng Y. Surface-engineered dendrimers in gene delivery. Chemical Reviews. 2015;**115**:5274-5300. DOI: 10.1021/cr500542t

[74] Muddineti OS, Ghosh B, Biswas S. Current trends in using polymer coated gold nanoparticles for cancer therapy. International Journal of Pharmaceutics. 2015;**484**:252-267. DOI: 10.1016/j.ijpharm.2015.02.038

[75] Lee J, Chatterjee DK, Lee MH, Krishnan S. Gold nanoparticles in breast cancer treatment: Promise and potential pitfalls. Cancer Letters. 2014;**347**:46-53. DOI: 10.1016/j.canlet.2014.02.006

[76] Nagy-Simon T, Tatar A-S, Craciun A-M, Vulpoi A, Jurj M-A, Florea A, et al. Antibody conjugated, Raman tagged hollow gold–silver nanospheres for specific targeting and multimodal dark-field/SERS/two photon-FLIM imaging of CD19(+) B lymphoblasts. ACS Applied Materials & Interfaces. 2017;**9**:21155-21168. DOI: 10.1021/acsami.7b05145

[77] Azzouzi S, Rotariu L, Benito AM, Maser WK, Ben Ali M, Bala C. A novel amperometric biosensor based on gold nanoparticles anchored on reduced graphene oxide for sensitive detection of l-lactate tumor biomarker. Biosensors & Bioelectronics. 2015;**69**:280-286. DOI: 10.1016/j.bios.2015.03.012

[78] Sun I-C, Na JH, Jeong SY, Kim D-E, Kwon IC, Choi K, et al. Biocompatible glycol chitosan-coated gold nanoparticles for tumor-targeting CT imaging. Pharmaceutical Research. 2014;**31**:1418-1425. DOI: 10.1007/s11095-013-1142-0

[79] Rengan AK, Bukhari AB, Pradhan A, Malhotra R, Banerjee R, Srivastava R, et al. In vivo analysis of biodegradable liposome gold nanoparticles as efficient agents for photothermal therapy of cancer. Nano Letters. 2015;**15**:842-848. DOI: 10.1021/nl5045378

[80] Wei L, Lu J, Xu H, Patel A, Chen Z-S, Chen G. Silver nanoparticles: Synthesis, properties, and therapeutic applications. Drug Discovery Today. 2015;**20**:595-601. DOI: 10.1016/j.drudis.2014.11.014

[81] Biswas A, Wang T, Biris AS. Single metal nanoparticle spectroscopy: Optical characterization of individual nanosystems for biomedical applications. Nanoscale. 2010;**2**:1560. DOI: 10.1039/c0nr00133c

[82] Zhang Y, Huang R, Zhu X, Wang L, Wu C. Synthesis, properties, and optical applications of noble metal nanoparticle-biomolecule conjugates. Chinese Science Bulletin. 2012;**57**:238-246. DOI: 10.1007/s11434-011-4747-x

[83] Abou El-Nour KMM, Eftaiha A, Al-Warthan A, Ammar RAA. Synthesis and applications of silver nanoparticles. Arabian Journal of Chemistry. 2010;**3**:135-140. DOI: 10.1016/J.ARABJC.2010.04.008

[84] Fernando S, Gunasekara T, Holton J. Antimicrobial nanoparticles: Applications and mechanisms of action. Sri Lankan Journal of Infectious Diseases. 2018;**8**:2. DOI: 10.4038/sljid.v8i1.8167

[85] Numnuam A, Thavarungkul P, Kanatharana P. An amperometric uric acid biosensor based on chitosan-carbon

nanotubes electrospun nanofiber on silver nanoparticles. Analytical and Bioanalytical Chemistry. 2014;**406**:3763-3772. DOI: 10.1007/s00216-014-7770-3

[86] Plackal Adimuriyil George B, Kumar N, Abrahamse H, Ray SS. Apoptotic efficacy of multifaceted biosynthesized silver nanoparticles on human adenocarcinoma cells. Scientific Reports. 2018;**8**:14368. DOI: 10.1038/s41598-018-32480-5

[87] He X, Gao J, Gambhir SS, Cheng Z. Near-infrared fluorescent nanoprobes for cancer molecular imaging: Status and challenges. Trends in Molecular Medicine. 2010;**16**:574-583. DOI: 10.1016/j.molmed.2010.08.006

[88] Medarova Z, Pham W, Farrar C, Petkova V, Moore A. In vivo imaging of siRNA delivery and silencing in tumors. Nature Medicine. 2007;**13**:372-377. DOI: 10.1038/nm1486

[89] Rhyner MN, Smith AM, Gao X, Mao H, Yang L, Nie S. Quantum dots and multifunctional nanoparticles: New contrast agents for tumor imaging. Nanomedicine. 2006;**1**:209-217. DOI: 10.2217/17435889.1.2.209

[90] Karimi Z, Karimi L, Shokrollahi H. Nano-magnetic particles used in biomedicine: Core and coating materials. Materials Science and Engineering: C. 2013;**33**:2465-2475. DOI: 10.1016/j.msec.2013.01.045

[91] Tural B, Özkan N, Volkan M. Preparation and characterization of polymer coated superparamagnetic magnetite nanoparticle agglomerates. Journal of Physics and Chemistry of Solids. 2009;**70**:860-866. DOI: 10.1016/J.JPCS.2009.04.007

[92] Sahu NK, Gupta J, Bahadur D. PEGylated FePt–Fe_3O_4 composite nanoassemblies (CNAs): in vitro hyperthermia, drug delivery and generation of reactive oxygen species (ROS). Dalton Transactions. 2015;**44**:9103-9113. DOI: 10.1039/C4DT03470H

[93] Ye F, Laurent S, Fornara A, Astolfi L, Qin J, Roch A, et al. Uniform mesoporous silica coated iron oxide nanoparticles as a highly efficient, nontoxic MRI T_2 contrast agent with tunable proton relaxivities. Contrast Media & Molecular Imaging. 2012;**7**:460-468. DOI: 10.1002/cmmi.1473

[94] Barick KC, Singh S, Bahadur D, Lawande MA, Patkar DP, Hassan PA. Carboxyl decorated Fe_3O_4 nanoparticles for MRI diagnosis and localized hyperthermia. Journal of Colloid and Interface Science. 2014;**418**:120-125. DOI: 10.1016/j.jcis.2013.11.076

[95] Topel SD, Topel Ö, Bostancıoğlu RB, Koparal AT. Synthesis and characterization of Bodipy functionalized magnetic iron oxide nanoparticles for potential bioimaging applications. Colloids and Surfaces. B, Biointerfaces. 2015;**128**:245-253. DOI: 10.1016/j.colsurfb.2015.01.043

[96] Zhao G, Wang J, Peng X, Li Y, Yuan X, Ma Y. Facile solvothermal synthesis of mesostructured Fe3O4/chitosan nanoparticles as delivery vehicles for pH-responsive drug delivery and magnetic resonance imaging contrast agents. Chemistry—An Asian Journal. 2014;**9**:546-553. DOI: 10.1002/asia.201301072

[97] Astefanei A, Núñez O, Galceran MT. Characterisation and determination of fullerenes: A critical review. Analytica Chimica Acta. 2015;**882**:1-21. DOI: 10.1016/j.aca.2015.03.025

[98] Chistyakov VA, Smirnova YO, Prazdnova EV, Soldatov AV. Possible

mechanisms of fullerene C 60 antioxidant action. BioMed Research International. 2013;**2013**:1-4. DOI: 10.1155/2013/821498

[99] Martinez ZS, Castro E, Seong C-S, Cerón MR, Echegoyen L, Llano M. Fullerene derivatives strongly inhibit HIV-1 replication by affecting virus maturation without impairing protease activity. Antimicrobial Agents and Chemotherapy. 2016;**60**:5731-5741. DOI: 10.1128/AAC.00341-16

[100] Shetti NP, Malode SJ, Nandibewoor ST. Electrochemical behavior of an antiviral drug acyclovir at fullerene-C60-modified glassy carbon electrode. Bioelectrochemistry. 2012;**88**:76-83. DOI: 10.1016/j.bioelechem.2012.06.004

[101] Bolskar RD. Fullerenes for Drug Delivery. In: Encyclopedia of Nanotechnology. Dordrecht: Springer Netherlands; 2016. pp. 1267-1281

[102] Kumar M, Raza K. C60-fullerenes as drug delivery carriers for anticancer agents: Promises and hurdles. Pharmaceutical Nanotechnology. 2018;**5**:169-179. DOI: 10.2174/2211738505666170301142232

[103] Maeda-Mamiya R, Noiri E, Isobe H, Nakanishi W, Okamoto K, Doi K, et al. In vivo gene delivery by cationic tetraamino fullerene. Proceedings of the National Academy of Sciences. 2010;**107**:5339-5344. DOI: 10.1073/pnas.0909223107

[104] Hamblin MR. Fullerenes as photosensitizers in photodynamic therapy: Pros and cons. Photochemical & Photobiological Sciences. 2018;**17**:1515-1533. DOI: 10.1039/C8PP00195B

[105] Grebinyk A, Grebinyk S, Prylutska S, Ritter U, Matyshevska O, Dandekar T, et al. C60 fullerene accumulation in human leukemic cells and perspectives of LED-mediated photodynamic therapy. Free Radical Biology & Medicine. 2018;**124**:319-327. DOI: 10.1016/j.freeradbiomed.2018.06.022

[106] Ibrahim KS. Carbon nanotubes-properties and applications: A review. Carbon Letters. 2013;**14**:131-144. DOI: 10.5714/CL.2013.14.3.131

[107] Chen Z, Zhang A, Wang X, Zhu J, Fan Y, Yu H, et al. The advances of carbon nanotubes in cancer diagnostics and therapeutics. Journal of Nanomaterials. 2017;**2017**:1-13. DOI: 10.1155/2017/3418932

[108] Sanginario A, Miccoli B, Demarchi D. Carbon nanotubes as an effective opportunity for cancer diagnosis and treatment. Biosensors. 2017;7:9. DOI: 10.3390/bios7010009

[109] Hong H, Gao T, Cai W. Molecular imaging with single-walled carbon nanotubes. Nano Today. 2009;**4**:252-261. DOI: 10.1016/j.nantod.2009.04.002

[110] Liu Z, Tabakman S, Sherlock S, Li X, Chen Z, Jiang K, et al. Multiplexed five-color molecular imaging of cancer cells and tumor tissues with carbon nanotube Raman tags in the near-infrared. Nano Research. 2010;**3**:222-233. DOI: 10.1007/s12274-010-1025-1

[111] Guo Q, Shen X, Li Y, Xu S. Carbon nanotubes-based drug delivery to cancer and brain. Current Medical Science. 2017;**37**:635-641. DOI: 10.1007/s11596-017-1783-z

[112] Madani S Y, Naderi N, Dissanayake O, Tan A, Seifalian A M. A new era of cancer treatment: Carbon nanotubes as drug delivery tools. International Journal of Nanomedicine. 2011;**6**:2963-2979. DOI: 10.2147/IJN.S16923

[113] Son KH, Hong JH, Lee JW. Carbon nanotubes as cancer therapeutic carriers

and mediators. International Journal of Nanomedicine. 2016;**11**:5163-5185. DOI: 10.2147/IJN.S112660

[114] Sobhani Z, Behnam MA, Emami F, Dehghanian A, Jamhiri I. Photothermal therapy of melanoma tumor using multiwalled carbon nanotubes. International Journal of Nanomedicine. 2017;**12**:4509-4517. DOI: 10.2147/IJN.S134661

[115] Eldridge BN, Bernish BW, Fahrenholtz CD, Singh R. Photothermal therapy of glioblastoma multiforme using multiwalled carbon nanotubes optimized for diffusion in extracellular space. ACS Biomaterials Science & Engineering. 2016;**2**:963-976. DOI: 10.1021/acsbiomaterials.6b00052

[116] Sigmund W, Yuh J, Park H, Maneeratana V, Pyrgiotakis G, Daga A, et al. Processing and structure relationships in electrospinning of ceramic fiber systems. Journal of the American Ceramic Society. 2006;**89**:395-407. DOI: 10.1111/j.1551-2916.2005.00807.x

[117] Thomas SC, Harshita, Mishra PK, Talegaonkar S. Ceramic nanoparticles: Fabrication methods and applications in drug delivery. Current Pharmaceutical Design. 2015;**21**:6165-6188

[118] Kim J, Jo C, Lim W-G, Jung S, Lee YM, Lim J, et al. Programmed nanoparticle-loaded nanoparticles for deep-penetrating 3D cancer therapy. Advanced Materials. 2018;**30**:1707557. DOI: 10.1002/adma.201707557

[119] Li S, Jiang Q, Liu S, Zhang Y, Tian Y, Song C, et al. A DNA nanorobot functions as a cancer therapeutic in response to a molecular trigger in vivo. Nature Biotechnology. 2018;**36**:258-264. DOI: 10.1038/nbt.4071

[120] Ahn S, Seo E, Kim K, Lee SJ. Controlled cellular uptake and drug efficacy of nanotherapeutics. Scientific Reports. 2013;**3**:1997. DOI: 10.1038/srep01997

[121] Richards DM, Endres RG. The mechanism of phagocytosis: Two stages of engulfment. Biophysical Journal. 2014;**107**:1542-1553. DOI: 10.1016/j.bpj.2014.07.070

[122] Rosales C, Uribe-Querol E. Phagocytosis: A fundamental process in immunity. BioMed Research International. 2017;**2017**:9042851. DOI: 10.1155/2017/9042851

[123] Bloomfield G, Kay RR. Uses and abuses of macropinocytosis. Journal of Cell Science. 2016;**129**:2697-2705. DOI: 10.1242/jcs.176149

[124] Buckley CM, King JS. Drinking problems: Mechanisms of macropinosome formation and maturation. The FEBS Journal. 2017;**284**:3778-3790. DOI: 10.1111/febs.14115

[125] Lim JP, Gleeson PA. Macropinocytosis: An endocytic pathway for internalising large gulps. Immunology and Cell Biology. 2011;**89**:836-843. DOI: 10.1038/icb.2011.20

[126] Lajoie P, Nabi IR. Lipid rafts, caveolae, and their endocytosis. International Review of Cell and Molecular Biology. 2010;**282**:135-163

[127] Pelkmans L, Helenius A. Endocytosis via caveolae. Traffic. 2002;**3**:311-320

[128] Kirkham M, Parton RG. Clathrin-independent endocytosis: New insights into caveolae and non-caveolar lipid raft carriers. Biochimica et Biophysica Acta, Molecular Cell Research. 2005;**1745**:273-286. DOI: 10.1016/j.bbamcr.2005.06.002

[129] Chen K, Li X, Zhu H, Gong Q, Luo K. Endocytosis of nanoscale

systems for cancer treatments. Current Medicinal Chemistry. 2018;**25**:3017-3035. DOI: 10.2174/0929867324666170428153056

[130] Ferguson JP, Huber SD, Willy NM, Aygün E, Goker S, Atabey T, et al. Mechanoregulation of clathrin-mediated endocytosis. Journal of Cell Science. 2017;**130**:3631-3636. DOI: 10.1242/jcs.205930

[131] O'Kelly I. Endocytosis as a mode to regulate functional expression of two-pore domain potassium (K2P) channels. Pflügers Archiv—European Journal of Physiology. 2015;**467**:1133-1142. DOI: 10.1007/s00424-014-1641-9

[132] Mayor S, Pagano RE. Pathways of clathrin-independent endocytosis. Nature Reviews Molecular Cell Biology. 2007;**8**:603-612. DOI: 10.1038/nrm2216

[133] Petros RA, DeSimone JM. Strategies in the design of nanoparticles for therapeutic applications. Nature Reviews Drug Discovery. 2010;**9**:615-627. DOI: 10.1038/nrd2591

[134] Sharma A, Vaghasiya K, Ray E, Verma RK. Lysosomal targeting strategies for design and delivery of bioactive for therapeutic interventions. Journal of Drug Targeting. 2018;**26**:208-221. DOI: 10.1080/1061186X.2017.1374390

[135] Seveau S. Multifaceted Activity of Listeriolysin O, the Cholesterol-Dependent Cytolysin of *Listeria monocytogenes*. Dordrecht: Springer; 2014. pp. 161-195

[136] Yu H, Zou Y, Wang Y, Huang X, Huang G, Sumer BD, et al. Overcoming endosomal barrier by amphotericin B-loaded dual pH-responsive PDMA-*b*-PDPA micelleplexes for siRNA delivery. ACS Nano. 2011;**5**:9246-9255. DOI: 10.1021/nn203503h

[137] Hou KK, Pan H, Schlesinger PH, Wickline SA. A role for peptides in overcoming endosomal entrapment in siRNA delivery—A focus on melittin. Biotechnology Advances. 2015;**33**:931-940. DOI: 10.1016/j.biotechadv.2015.05.005

[138] Nishimura Y, Takeda K, Ezawa R, Ishii J, Ogino C, Kondo A. A display of pH-sensitive fusogenic GALA peptide facilitates endosomal escape from a bio-nanocapsule via an endocytic uptake pathway. Journal of Nanobiotechnology. 2014;**12**:11. DOI: 10.1186/1477-3155-12-11

[139] Nakase I, Futaki S. Combined treatment with a pH-sensitive fusogenic peptide and cationic lipids achieves enhanced cytosolic delivery of exosomes. Scientific Reports. 2015;**5**:10112. DOI: 10.1038/srep10112

[140] Erazo-Oliveras A, Muthukrishnan N, Baker R, Wang T-Y, Pellois J-P. Improving the endosomal escape of cell-penetrating peptides and their cargos: Strategies and challenges. Pharmaceuticals (Basel). 2012;**5**:1177-1209. DOI: 10.3390/ph5111177

[141] Ohtsuki T, Miki S, Kobayashi S, Haraguchi T, Nakata E, Hirakawa K, et al. The molecular mechanism of photochemical internalization of cell penetrating peptide-cargo-photosensitizer conjugates. Scientific Reports. 2016;**5**:18577. DOI: 10.1038/srep18577

[142] Yang X, Fan B, Gao W, Li L, Li T, Sun J, et al. Enhanced endosomal escape by photothermal activation for improved small interfering RNA delivery and antitumor effect. International Journal of Nanomedicine. 2018;**13**:4333-4344. DOI: 10.2147/IJN.S161908

[143] Rollo FD. Molecular imaging: An overview and clinical applications. Radiology Management. 2003;**25**:28-32; quiz 33-35

[144] Jokerst JV, Gambhir SS. Molecular imaging with theranostic nanoparticles.

Accounts of Chemical Research. 2011;**44**:1050-1060. DOI: 10.1021/ar200106e

[145] Debbage P, Jaschke W. Molecular imaging with nanoparticles: Giant roles for dwarf actors. Histochemistry and Cell Biology. 2008;**130**:845-875. DOI: 10.1007/s00418-008-0511-y

[146] Kircher MF, Willmann JK. Molecular body imaging: MR imaging, CT, and US. Part I. Principles. Radiology. 2012;**263**:633-643. DOI: 10.1148/radiol.12102394

[147] van Beek EJR, Hoffman EA. Functional imaging: CT and MRI. Clinics in Chest Medicine. 2008;**29**:195-216. DOI: 10.1016/j.ccm.2007.12.003

[148] Kiessling F, Huppert J, Palmowski M. Functional and molecular ultrasound imaging: Concepts and contrast agents. Current Medicinal Chemistry. 2009;**16**:627-642. DOI: 10.2174/092986709787458470

[149] Gambhir SS, Czernin J, Schwimmer J, Silverman DH, Coleman RE, Phelps ME. A tabulated summary of the FDG PET literature. Journal of Nuclear Medicine. 2001;**42**:1S-93S

[150] Griffeth LK. Use of PET/CT scanning in cancer patients: Technical and practical considerations. Proceedings (Baylor University Medical Center). 2005;**18**:321-330

[151] Jadvar H, Colletti PM. Competitive advantage of PET/MRI. European Journal of Radiology. 2014;**83**:84-94. DOI: 10.1016/j.ejrad.2013.05.028

[152] Khalil MM, Tremoleda JL, Bayomy TB, Gsell W. Molecular SPECT imaging: An overview. International Journal of Molecular Imaging. 2011;**2011**:796025. DOI: 10.1155/2011/796025

[153] Gnanasegaran G, Ballinger JR. Molecular imaging agents for SPECT (and SPECT/CT). European Journal of Nuclear Medicine and Molecular Imaging. 2014;**41**:26-35. DOI: 10.1007/s00259-013-2643-0

[154] Schmidt CW. CT scans: Balancing health risks and medical benefits. Environmental Health Perspectives. 2012;**120**:A118-A121. DOI: 10.1289/ehp.120-a118

[155] Hill MA, O'Neill P, McKenna WG. Comments on potential health effects of MRI-induced DNA lesions: Quality is more important to consider than quantity. European Heart Journal Cardiovascular Imaging. 2016;**17**:1230-1238. DOI: 10.1093/ehjci/jew163

[156] Hladik WB, Norenberg JP. Problems associated with the clinical use of radiopharmaceuticals: A proposed classification system and troubleshooting guide. European Journal of Nuclear Medicine. 1996;**23**:997-1002

[157] Shukla A, Kumar U. Positron emission tomography: An overview. Journal of Medical Physics. 2006;**31**:13. DOI: 10.4103/0971-6203.25665

[158] Huang Y-Y. An Overview of PET radiopharmaceuticals in clinical use: Regulatory, quality and pharmacopeia monographs of the United States and Europe. In: Nuclear Medicine Physics. Rijeka, Croatia: IntechOpen; 2018

[159] Shankar H, Pagel PS. Potential adverse ultrasound-related biological effects. Anesthesiology. 2011;**115**:1109-1124. DOI: 10.1097/ALN.0b013e31822fd1f1

[160] Anani T, Panizzi P, David AE. Nanoparticle-based probes to enable noninvasive imaging of proteolytic activity for cancer diagnosis. Nanomedicine. 2016;**11**:2007-2022. DOI: 10.2217/nnm-2016-0027

[161] Heneweer C, Grimm J. Clinical applications in molecular imaging. Pediatric Radiology. 2011;**41**:199-207. DOI: 10.1007/s00247-010-1902-5

[162] Boschi F, De Sanctis F. Overview of the optical properties of fluorescent nanoparticles for optical imaging. European Journal of Histochemistry. 2017;**61**:2830. DOI: 10.4081/ejh.2017.2830

[163] Daniel M, Astruc D. Gold nanoparticles: Assembly, supramolecular chemistry, quantum-size-related properties, and applications toward biology, catalysis, and nanotechnology. Chemical Reviews. 2004;**104**:293-346. DOI: 10.1021/CR030698+

[164] Choi J, Shin D-M, Song H, Lee D, Kim K. Current achievements of nanoparticle applications in developing optical sensing and imaging techniques. Nano Convergence. 2016;**3**:30. DOI: 10.1186/s40580-016-0090-x

[165] Ștefan N, Moldovan AI, Toma V, Moldovan CS, Berindan-Neagoe I, Știufiuc G, et al. PEGylated gold nanoparticles with interesting plasmonic properties synthesized using an original, rapid, and easy-to-implement procedure. Journal of Nanomaterials. 2018;**2018**:1-7. DOI: 10.1155/2018/5954028

[166] Hu K, Wang H, Tang G, Huang T, Tang X, Liang X, et al. In vivo cancer dual-targeting and dual-modality imaging with functionalized quantum dots. Journal of Nuclear Medicine. 2015;**56**:1278-1284. DOI: 10.2967/jnumed.115.158873

[167] Sun I-C, Eun D-K, Koo H, Ko C-Y, Kim H-S, Yi DK, et al. Tumor-targeting gold particles for dual computed tomography/optical cancer imaging. Angewandte Chemie, International Edition. 2011;**50**:9348-9351. DOI: 10.1002/anie.201102892

[168] Popovtzer R, Agrawal A, Kotov NA, Popovtzer A, Balter J, Carey TE, et al. Targeted gold nanoparticles enable molecular CT imaging of cancer. Nano Letters. 2008;**8**:4593-4596

[169] Black KCL, Akers WJ, Sudlow G, Xu B, Laforest R, Achilefu S. Dual-radiolabeled nanoparticle SPECT probes for bioimaging. Nanoscale. 2015;**7**:440-444. DOI: 10.1039/C4NR05269B

[170] Chen T-J, Cheng T-H, Chen C-Y, Hsu SCN, Cheng T-L, Liu G-C, et al. Targeted herceptin–dextran iron oxide nanoparticles for noninvasive imaging of HER2/neu receptors using MRI. JBIC, Journal of Biological Inorganic Chemistry. 2009;**14**:253-260. DOI: 10.1007/s00775-008-0445-9

[171] Gallo J, Kamaly N, Lavdas I, Stevens E, Nguyen Q-D, Wylezinska-Arridge M, et al. CXCR4-targeted and MMP-responsive iron oxide nanoparticles for enhanced magnetic resonance imaging. Angewandte Chemie, International Edition. 2014;**53**:9550-9554. DOI: 10.1002/anie.201405442

[172] Chen F, Hong H, Shi S, Goel S, Valdovinos HF, Hernandez R, et al. Engineering of hollow mesoporous silica nanoparticles for remarkably enhanced tumor active targeting efficacy. Scientific Reports. 2015;**4**:5080. DOI: 10.1038/srep05080

[173] Hu Z, Yang B, Li T, Li J. Thyroid cancer detection by ultrasound molecular imaging with SHP2-targeted perfluorocarbon nanoparticles. Contrast Media & Molecular Imaging. 2018;**2018**:1-7. DOI: 10.1155/2018/8710862

[174] Liu J, Li J, Rosol TJ, Pan X, Voorhees JL. Biodegradable nanoparticles for targeted ultrasound imaging of breast cancer cells *in vitro*. Physics in Medicine and Biology. 2007;**52**:4739-4747. DOI: 10.1088/0031-9155/52/16/002

[175] Barrett T, Ravizzini G, Choyke P, Kobayashi H. Dendrimers in medical nanotechnology. IEEE Engineering in Medicine and Biology Magazine. 2009;**28**:12-22. DOI: 10.1109/MEMB.2008.931012

[176] Zhu C, Liu L, Yang Q, Lv F, Wang S. Water-soluble conjugated polymers for imaging, diagnosis, and therapy. Chemical Reviews. 2012;**112**:4687-4735. DOI: 10.1021/cr200263w

[177] Wang T, Ray J. Aptamer-based molecular imaging. Protein & Cell. 2012;**3**:739-754. DOI: 10.1007/s13238-012-2072-z

[178] Chen S, Yu Y-L, Wang J-H. Inner filter effect-based fluorescent sensing systems: A review. Analytica Chimica Acta. 2018;**999**:13-26. DOI: 10.1016/J.ACA.2017.10.026

[179] Zhang H, Zhang B, Di C, Ali MC, Chen J, Li Z, et al. Label-free fluorescence imaging of cytochrome c in living systems and anti-cancer drug screening with nitrogen doped carbon quantum dots. Nanoscale. 2018;**10**:5342-5349. DOI: 10.1039/c7nr08987b

[180] Zhang F, He X, Ma P, Sun Y, Wang X, Song D. Rapid aqueous synthesis of CuInS/ZnS quantum dots as sensor probe for alkaline phosphatase detection and targeted imaging in cancer cells. Talanta. 2018;**189**:411-417. DOI: 10.1016/j.talanta.2018.07.031

[181] Luo J, Xie Z, Lam JW, Cheng L, Chen H, Qiu C, et al. Aggregation-induced emission of 1-methyl-1,2,3,4,5-pentaphenylsilole. Chemical Communications. 2001;**18**:1740-1741

[182] Wang D, Lee MMS, Xu W, Kwok RTK, Lam JWY, Tang BZ. Theranostics based on AIEgens. Theranostics. 2018;**8**:4925-4956. DOI: 10.7150/thno.27787

[183] Yuan Y, Kwok RTK, Tang BZ, Liu B. Targeted theranostic platinum(IV) prodrug with a built-in aggregation-induced emission light-up apoptosis sensor for noninvasive early evaluation of its therapeutic responses in situ. Journal of the American Chemical Society. 2014;**136**:2546-2554. DOI: 10.1021/ja411811w

[184] Din FU, Aman W, Ullah I, Qureshi OS, Mustapha O, Shafique S, et al. Effective use of nanocarriers as drug delivery systems for the treatment of selected tumors. International Journal of Nanomedicine. 2017;**12**:7291-7309. DOI: 10.2147/IJN.S146315

[185] Prasad D, Chauhan H. Key targeting approaches for pharmaceutical drug delivery. American Pharmaceutical Review. 2013;**16**:6

[186] Shi S, Kong N, Feng C, Shajii A, Bejgrowicz C, Tao W, et al. Drug delivery strategies for the treatment of metabolic diseases. Advanced Healthcare Materials. 2019;**1801655**:1801655. DOI: 10.1002/adhm.201801655

[187] Joshi D, Garg T, Goyal AK, Rath G. Advanced drug delivery approaches against periodontitis. Drug Delivery. 2016;**23**:363-377. DOI: 10.3109/10717544.2014.935531

[188] Sarigol-Calamak E, Hascicek C. Tissue scaffolds as a local drug delivery system for bone regeneration. Advances in Experimental Medicine and Biology. 2018;**1078**:475-493. DOI: 10.1007/978-981-13-0950-2_25

[189] MaedaH, MatsumuraY. Tumoritropic and lymphotropic principles of macromolecular drugs. Critical Reviews in Therapeutic Drug Carrier Systems. 1989;**6**:193-210

[190] Matsumura Y, Maeda H. A new concept for macromolecular therapeutics in cancer chemotherapy:

Mechanism of tumoritropic accumulation of proteins and the antitumor agent smancs. Cancer Research. 1986;**46**:6387-6392

[191] Bazak R, Houri M, El AS, Hussein W, Refaat T. Passive targeting of nanoparticles to cancer: A comprehensive review of the literature. Molecular and Clinical Oncology. 2014;**2**:904-908. DOI: 10.3892/mco.2014.356

[192] Yu X, Trase I, Ren M, Duval K, Guo X, Chen Z. Design of nanoparticle-based carriers for targeted drug delivery. Journal of Nanomaterials. 2016;**2016**:1-15. DOI: 10.1155/2016/1087250

[193] Karimi M, Eslami M, Sahandi-Zangabad P, Mirab F, Farajisafiloo N, Shafaei Z, et al. pH-Sensitive stimulus-responsive nanocarriers for targeted delivery of therapeutic agents. Wiley Interdisciplinary Reviews. Nanomedicine and Nanobiotechnology. 2016;**8**:696-716. DOI: 10.1002/wnan.1389

[194] Polyak B, Friedman G. Magnetic targeting for site-specific drug delivery: Applications and clinical potential. Expert Opinion on Drug Delivery. 2009;**6**:53-70. DOI: 10.1517/17425240802662795

[195] Rukshin I, Mohrenweiser J, Yue P, Afkhami S. Modeling superparamagnetic particles in blood flow for applications in magnetic drug targeting. Fluids. 2017;**2**:29. DOI: 10.3390/fluids2020029

[196] Yoo J, Park C, Yi G, Lee D, Koo H, Yoo J, et al. Active targeting strategies using biological ligands for nanoparticle drug delivery systems. Cancers. 2019;**11**:640. DOI: 10.3390/cancers11050640

[197] Tokunaga S, Takashima T, Kashiwagi S, Noda S, Kawajiri H, Tokumoto M, et al. Neoadjuvant chemotherapy with nab-paclitaxel plus trastuzumab followed by 5-fluorouracil/epirubicin/cyclophosphamide for HER2-positive operable breast cancer: A multicenter phase II trial. Anticancer Research. 2019;**39**:2053-2059. DOI: 10.21873/anticanres.13316

[198] Safran H, DiPetrillo T, Akerman P, Ng T, Evans D, Steinhoff M, et al. Phase I/II study of trastuzumab, paclitaxel, cisplatin and radiation for locally advanced, HER2 overexpressing, esophageal adenocarcinoma. International Journal of Radiation Oncology. 2007;**67**:405-409. DOI: 10.1016/j.ijrobp.2006.08.076

[199] Tummers WS, Miller SE, Teraphongphom NT, Gomez A, Steinberg I, Huland DM, et al. Intraoperative pancreatic cancer detection using tumor-specific multimodality molecular imaging. Annals of Surgical Oncology. 2018;**25**:1880-1888. DOI: 10.1245/s10434-018-6453-2

[200] Rosenthal EL, Kulbersh BD, King T, Chaudhuri TR, Zinn KR. Use of fluorescent labeled anti-epidermal growth factor receptor antibody to image head and neck squamous cell carcinoma xenografts. Molecular Cancer Therapeutics. 2007;**6**:1230-1238. DOI: 10.1158/1535-7163.MCT-06-0741

[201] Rosenthal EL, Kulbersh BD, Duncan RD, Zhang W, Magnuson JS, Carroll WR, et al. In vivo detection of head and neck cancer orthotopic xenografts by immunofluorescence. Laryngoscope. 2006;**116**:1636-1641. DOI: 10.1097/01.mlg.0000232513.19873.da

[202] Jiang B, Qi JY, Sun MY, Li ZJ, Liu W, Liu LJ, et al. Tolerance and pharmacodynamics phase I clinical trial study of chimeric anti-CD20 monoclonal antibody IBI301 in Chinese patients with CD20-positive non-Hodgkin's lymphoma.

Zhonghua Xue Ye Xue Za Zhi. 2018;**39**:320-324. DOI: 10.3760/cma.j.issn.0253-2727.2018.04.013

[203] Edwards JCW, Szczepański L, Szechiński J, Filipowicz-Sosnowska A, Emery P, Close DR, et al. Efficacy of B-cell–targeted therapy with rituximab in patients with rheumatoid arthritis. The New England Journal of Medicine. 2004;**350**:2572-2581. DOI: 10.1056/NEJMoa032534

[204] Luria-Pérez R, Helguera G, Rodríguez JA. Antibody-mediated targeting of the transferrin receptor in cancer cells. Boletín Médico del Hospital Infantil de México. 2016;**73**:372-379. DOI: 10.1016/j.bmhimx.2016.11.004

[205] Jhaveri A, Deshpande P, Pattni B, Torchilin V. Transferrin-targeted, resveratrol-loaded liposomes for the treatment of glioblastoma. Journal of Controlled Release. 2018;**277**:89-101. DOI: 10.1016/j.jconrel.2018.03.006

[206] Dhir V, Sandhu A, Kaur J, Pinto B, Kumar P, Kaur P, et al. Comparison of two different folic acid doses with methotrexate—A randomized controlled trial (FOLVARI study). Arthritis Research & Therapy. 2015;**17**:156. DOI: 10.1186/s13075-015-0668-4

[207] Ledermann JA, Canevari S, Thigpen T. Targeting the folate receptor: Diagnostic and therapeutic approaches to personalize cancer treatments. Annals of Oncology. 2015;**26**:2034-2043. DOI: 10.1093/annonc/mdv250

[208] Fan X, Guo Y, Wang L, Xiong X, Zhu L, Fang K. Diagnosis of prostate cancer using anti-PSMA aptamer A10-3.2-oriented lipid nanobubbles. International Journal of Nanomedicine. 2016;**11**:3939-3950. DOI: 10.2147/IJN.S112951

[209] Baneshi M, Dadfarnia S, Shabani AMH, Sabbagh SK, Haghgoo S, Bardania H. A novel theranostic system of AS1411 aptamer-functionalized albumin nanoparticles loaded on iron oxide and gold nanoparticles for doxorubicin delivery. International Journal of Pharmaceutics. 2019;**564**:145-152. DOI: 10.1016/j.ijpharm.2019.04.025

[210] Mohammadinejad A, Taghdisi SM, Es'haghi Z, Abnous K, Mohajeri SA. Targeted imaging of breast cancer cells using two different kinds of aptamers -functionalized nanoparticles. European Journal of Pharmaceutical Sciences. 2019;**134**:60-68. DOI: 10.1016/j.ejps.2019.04.012

[211] Andreou C, Pal S, Rotter L, Yang J, Kircher MF. Molecular imaging in nanotechnology and theranostics. Molecular Imaging and Biology. 2017;**19**:363-372. DOI: 10.1007/s11307-017-1056-z

[212] Yildiz T, Gu R, Zauscher S, Betancourt T. Doxorubicin-loaded protease-activated near-infrared fluorescent polymeric nanoparticles for imaging and therapy of cancer. International Journal of Nanomedicine. 2018;**13**:6961-6986. DOI: 10.2147/IJN.S174068

[213] Dobiasch S, Szanyi S, Kjaev A, Werner J, Strauss A, Weis C, et al. Synthesis and functionalization of protease-activated nanoparticles with tissue plasminogen activator peptides as targeting moiety and diagnostic tool for pancreatic cancer. Journal of Nanobiotechnology. 2016;**14**:81. DOI: 10.1186/s12951-016-0236-3

[214] Yigit MV, Moore A, Medarova Z. Magnetic nanoparticles for cancer diagnosis and therapy. Pharmaceutical Research. 2012;**29**:1180-1188. DOI: 10.1007/s11095-012-0679-7

[215] Lee S, Ryu JH, Park K, Lee A, Lee S-Y, Youn I-C, et al. Polymeric nanoparticle-based activatable

near-infrared nanosensor for protease determination in vivo. Nano Letters. 2009;**9**:4412-4416. DOI: 10.1021/nl902709m

[216] Zhang R, Lu W, Wen X, Huang M, Zhou M, Liang D, et al. Annexin A5-conjugated polymeric micelles for dual SPECT and optical detection of apoptosis. Journal of Nuclear Medicine. 2011;**52**:958-964. DOI: 10.2967/jnumed.110.083220

[217] Norregaard K, Jørgensen JT, Simón M, Melander F, Kristensen LK, Bendix PM, et al. 18F-FDG PET/CT-based early treatment response evaluation of nanoparticle-assisted photothermal cancer therapy. PLoS One. 2017;**12**:e0177997. DOI: 10.1371/journal.pone.0177997

[218] Loo C, Lowery A, Halas N, West J, Drezek R. Immunotargeted nanoshells for integrated cancer imaging and therapy. Nano Letters. 2005;**5**:709-711. DOI: 10.1021/nl050127s

[219] Rapoport N, Gao Z, Kennedy A. Multifunctional nanoparticles for combining ultrasonic tumor imaging and targeted chemotherapy. Journal of the National Cancer Institute. 2007;**99**:1095-1106. DOI: 10.1093/jnci/djm043

[220] Sorace AG, Warram JM, Umphrey H, Hoyt K. Microbubble-mediated ultrasonic techniques for improved chemotherapeutic delivery in cancer. Journal of Drug Targeting. 2012;**20**:43-54. DOI: 10.3109/1061186X.2011.622397

[221] Savla R, Taratula O, Garbuzenko O, Minko T. Tumor targeted quantum dot-mucin 1 aptamer-doxorubicin conjugate for imaging and treatment of cancer. Journal of Controlled Release. 2011;**153**:16-22. DOI: 10.1016/j.jconrel.2011.02.015

[222] Agostinis P, Berg K, Cengel KA, Foster TH, Girotti AW, Gollnick SO, et al. Photodynamic therapy of cancer: An update. CA: A Cancer Journal for Clinicians. 2011;**61**:250-281. DOI: 10.3322/caac.20114

[223] Yi G, Hong SH, Son J, Yoo J, Park C, Choi Y, et al. Recent advances in nanoparticle carriers for photodynamic therapy. Quantitative Imaging in Medicine and Surgery. 2018;**8**:433-443. DOI: 10.21037/qims.2018.05.04

[224] Sun Y, Campisi J, Higano C, Beer TM, Porter P, Coleman I, et al. Treatment-induced damage to the tumor microenvironment promotes prostate cancer therapy resistance through WNT16B. Nature Medicine. 2012;**18**:1359-1368. DOI: 10.1038/nm.2890

[225] Abrahamse H, Kruger CA, Kadanyo S, Mishra A. Nanoparticles for advanced photodynamic therapy of cancer. Photomedicine and Laser Surgery. 2017;**35**:581-588. DOI: 10.1089/pho.2017.4308

[226] Li W-T. Nanoparticles for photodynamic therapy. In: Handbook of Biophotonics. Weinheim, Germany: Wiley-VCH Verlag GmbH & Co. KGaA; 2013. pp. 321-336

[227] Calixto G, Bernegossi J, de Freitas L, Fontana C, Chorilli M. Nanotechnology-based drug delivery systems for photodynamic therapy of cancer: A review. Molecules. 2016;**21**:342. DOI: 10.3390/molecules21030342

[228] Mangadlao JD, Wang X, McCleese C, Escamilla M, Ramamurthy G, Wang Z, et al. Prostate-specific membrane antigen targeted gold nanoparticles for theranostics of prostate cancer. ACS Nano. 2018;**12**:3714-3725. DOI: 10.1021/acsnano.8b00940

[229] Li Y, Du Y, Liang X, Sun T, Xue H, Tian J, et al. EGFR-targeted liposomal nanohybrid cerasomes:

Theranostic function and immune checkpoint inhibition in a mouse model of colorectal cancer. Nanoscale. 2018;**10**:16738-16749. DOI: 10.1039/c8nr05803b

[230] Zou L, Wang H, He B, Zeng L, Tan T, Cao H, et al. Current approaches of photothermal therapy in treating cancer metastasis with nanotherapeutics. Theranostics. 2016;**6**:762-772. DOI: 10.7150/thno.14988

[231] Jaque D, Martínez Maestro L, del Rosal B, Haro-Gonzalez P, Benayas A, Plaza JL, et al. Nanoparticles for photothermal therapies. Nanoscale. 2014;**6**:9494-9530. DOI: 10.1039/c4nr00708e

[232] Lei T, Manchanda R, Fernandez-Fernandez A, Huang Y-C, Wright D, McGoron AJ. Thermal and pH sensitive multifunctional polymer nanoparticles for cancer imaging and therapy. RSC Advances. 2014;**4**: 17959-17968. DOI: 10.1039/C4RA01112K

[233] Zhu X, Feng W, Chang J, Tan Y-W, Li J, Chen M, et al. Temperature-feedback upconversion nanocomposite for accurate photothermal therapy at facile temperature. Nature Communications. 2016;7:10437. DOI: 10.1038/ncomms10437

[234] Liu S, Doughty A, West C, Tang Z, Zhou F, Chen WR. Determination of temperature distribution in tissue for interstitial cancer photothermal therapy. International Journal of Hyperthermia. 2018;**34**:756-763. DOI: 10.1080/02656736.2017.1370136

[235] Devalapally H, Shenoy D, Little S, Langer R, Amiji M. Poly(ethylene oxide)-modified poly(beta-amino ester) nanoparticles as a pH-sensitive system for tumor-targeted delivery of hydrophobic drugs: Part 3. Therapeutic efficacy and safety studies in ovarian cancer xenograft model. Cancer Chemotherapy and Pharmacology. 2007;**59**:477-484. DOI: 10.1007/s00280-006-0287-5

[236] Liong M, Lu J, Kovochich M, Xia T, Ruehm SG, Nel AE, et al. Multifunctional inorganic nanoparticles for imaging, targeting, and drug delivery. ACS Nano. 2008;**2**:889-896. DOI: 10.1021/nn800072t

[237] Hadjipanayis CG, Machaidze R, Kaluzova M, Wang L, Schuette AJ, Chen H, et al. EGFRvIII antibody-conjugated iron oxide nanoparticles for magnetic resonance imaging-guided convection-enhanced delivery and targeted therapy of glioblastoma. Cancer Research. 2010;**70**:6303-6312. DOI: 10.1158/0008-5472.CAN-10-1022

[238] von Maltzahn G, Park J-H, Agrawal A, Bandaru NK, Das SK, Sailor MJ, et al. Computationally guided photothermal tumor therapy using Long-circulating gold nanorod antennas. Cancer Research. 2009;**69**:3892-3900. DOI: 10.1158/0008-5472.CAN-08-4242

[239] Matea C, Mocan T, Tabaran F, Pop T, Mosteanu O, Puia C, et al. Quantum dots in imaging, drug delivery and sensor applications. International Journal of Nanomedicine. 2017;**12**:5421-5431. DOI: 10.2147/IJN.S138624

[240] Phillips E, Penate-Medina O, Zanzonico PB, Carvajal RD, Mohan P, Ye Y, et al. Clinical translation of an ultrasmall inorganic optical-PET imaging nanoparticle probe. Science Translational Medicine. 2014;**6**:260ra149-260ra149. DOI: 10.1126/scitranslmed.3009524

[241] Davis ME, Zuckerman JE, Choi CHJ, Seligson D, Tolcher A, Alabi CA, et al. Evidence of RNAi in humans from systemically administered siRNA via targeted nanoparticles. Nature. 2010;**464**:1067-1070. DOI: 10.1038/nature08956

[242] Singh P, Pandit S, Mokkapati VRSS, Garg A, Ravikumar V, Mijakovic I, et al. Gold nanoparticles in diagnostics and therapeutics for human cancer. International Journal of Molecular Sciences. 2018;**19**:1979. DOI: 10.3390/ijms19071979

[243] Libutti SK, Paciotti GF, Byrnes AA, Alexander HR, Gannon WE, Walker M, et al. Phase I and pharmacokinetic studies of CYT-6091, a novel PEGylated colloidal gold-rhTNF nanomedicine. Clinical Cancer Research. 2010;**16**: 6139-6149. DOI: 10.1158/1078-0432.CCR-10-0978

[244] Richards JMJ, Shaw CA, Lang NN, Williams MC, Semple SIK, MacGillivray TJ, et al. In vivo mononuclear cell tracking using superparamagnetic particles of iron oxide. Circulation. Cardiovascular Imaging. 2012;**5**:509-517. DOI: 10.1161/CIRCIMAGING.112.972596

[245] Yaari Z, da Silva D, Zinger A, Goldman E, Kajal A, Tshuva R, et al. Theranostic barcoded nanoparticles for personalized cancer medicine. Nature Communications. 2016;7:13325. DOI: 10.1038/ncomms13325

[246] Maniglio D, Benetti F, Minati L, Jovicich J, Valentini A, Speranza G, et al. Theranostic gold-magnetite hybrid nanoparticles for MRI-guided radiosensitization. Nanotechnology. 2018;**29**:315101. DOI: 10.1088/1361-6528/aac4ce

[247] Saliev T, Akhmetova A, Kulsharova G. Multifunctional hybrid nanoparticles for theranostics. In: Core-Shell Nanostructures Drug delivery and Theranostics: Challenges, Strategies and Prospects for Novel Carrier Systems. Sawston, Cambridge: Elsevier; 2018. pp. 177-244. DOI: 10.1016/B978-0-08-102198-9.00007-7

[248] Rajkumar S, Prabaharan M. Theranostic application of Fe_3O_4–Au hybrid nanoparticles. In: Noble Metal-Metal Oxide Hybrid Nanoparticles. Fundamentals and Applications. Sawston, Cambridge: Elsevier; 2019. pp. 607-623. DOI: 10.1016/B978-0-12-814134-2.00029-2

Chapter 8

Biological and Physical Applications of Silver Nanoparticles with Emerging Trends of Green Synthesis

Atamjit Singh and Kirandeep Kaur

Abstract

Among the emerging nanotechnology, nanoparticles get much attention due to their unique physicochemical, optical, electrical, and thermal activities. Nowadays, extensive research on silver nanoparticles is going on due to their wide applicability in different fields. Silver nanoparticles possess excellent anticancer as well as antimicrobial efficacy (hence found major and wide applications as antimicrobial, wound healing, antidiarrheal, and antifungal agents). A huge and advanced perspective of silver nanoparticles is found in environmental hygiene and sterilization due to their magnificent disinfectant properties. The other major applications of silver nanoparticles include diagnostic (as biological tags in biosensors, assays, and quantitative detection), conductive (in conductive inks, pastes, and fillers), optical (metal-enhanced fluorescence and surface-enhanced Raman scattering), and household (pesticides and wastewater treatment) applications. The present review consists of an exhaustive detail about the biological and physical applications of silver nanoparticles along with the analysis of historical evolution, the present scenario, and possible future outcomes.

Keywords: silver nanoparticles, anticancer, antimicrobial, environmental hygiene, biosensors

1. Introduction

In this modern era, pharmaceutical research associated with nano-sized products is rapidly growing. Nanoscience/technology has changed the way of diagnosing, treating, and curing the diseases which proves to be a great change in human life. Nano-sized formulations/products include nano-emulsion, ethosomes, liposomes, nanoparticles, etc. Nanoparticles ranging from 1 to 100 nm are in trend nowadays due to its size-depending optical, thermal, electrical, and biological properties [1]. Nano-sized metallic particles are unique because they can considerably change their chemical, physical, and biological properties because of their surface-to-volume ratio. Silver nanoparticles have unique physical and chemical properties among other metallic nanoparticles; besides this, its wide applications in different fields make them the most catchy and different from all other nano-formulations. Silver nanoparticles are well recognized for their diagnostic (as biological tags in biosensors, assays, and quantitative detection), conductive (in conductive inks, pastes, and fillers), optical (metal-enhanced fluorescence and surface-enhanced Raman scattering), and household (pesticides and wastewater treatment) applications. Silver nanoparticles gained their immense attraction due to its magnificent role in cancer treatment. The biological activity of silver nanoparticles depends upon various factors like surface morphology, surface chemistry, size, size distribution, cell type, cell agglomeration, and reducing agent used for the synthesis of nanoparticles. Silver nanoparticles were firstly recorded by M.C. Lea; by citrate reduction method, he produced stabilized silver colloids. Many methods are there for the synthesis of silver nanoparticle which include a physical method, chemical method, biological method, etc. Physical and chemical methods are somewhat hazardous and costly, whereas biological methods are safe and are simpler to apply for the synthesis of silver nanoparticles. After synthesis and before applying it for any purpose, silver nanoparticles must pass all the characteristic parameters like size, shape, size distribution, surface area, solubility, aggregation, toxicity, and biocompatibility. Many techniques have been used to evaluate all these parameters like UV-Vis spectroscopy, differential screening calorimetry (DSC), X-ray diffraction (XRD), Fourier transform infrared spectroscopy (FTIR), X-ray photoelectron spectroscopy (XPS), dynamic light scattering (DLS), scanning electron microscopy (SEM), transmission electron microscopy (TEM), and atomic force microscopy (AFM) [2–6].

Advantages of silver nanoparticles [7]:

- There is a possibility of high-scale production of silver nanoparticles.
- Silver nanoparticles possess long-term stability.
- Controlled drug delivery of silver nanoparticles can be achieved.
- Silver nanoparticles can be freeze-dried and lyophilized to get powder formulation.

Disadvantages of silver nanoparticles [7]:

- Less drug loading capacity.
- Dispersion of silver nanoparticles includes some amount of water.
- The less capacity to load lipophobic drugs.

2. Methods of preparation

2.1 Physical methods

Physical methods use physical energies to produce the silver nanoparticles with narrow size distribution. Physical methods produce a large quantity of silver nanoparticles in a single process. These methods are also able to give silver nanoparticle powder (**Figure 1**) [8].

2.1.1 Evaporation-condensation method

In this method, the metallic (silver-organic) source is kept in the boat with the heat center in a tube furnace. Center heat is enough to evaporate the non-silver

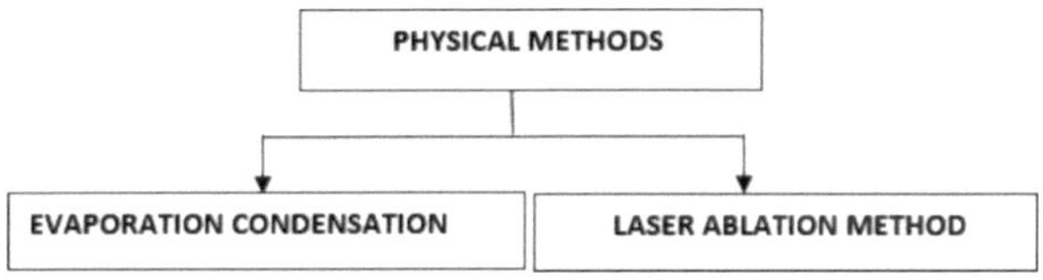

Figure 1.
Physical methods for the preparation of silver nanoparticles.

particles which get eliminated with the carrier gas leaving behind the silver nanoparticles. The more the temperature of the furnace, the more the concentration of silver nanoparticles formed. But this method takes a quite large time to reach stabilized temperature [9].

2.1.2 Laser ablation method

In this method, metallic/silver plate is dispersed in a liquid medium and illuminated with a laser beam. The metal plate absorbs the laser beam and forms a hot plasma which contains silver particles in maximum concentration. The liquid medium lowers down the temperature and cools the vicinity which initiates the formation of silver nanoparticles. The nature of the silver nanoparticles formed and the ablation efficiency depends upon many factors such as the wavelength of the laser impinging the metallic target, the duration of the laser pulses (in the femto-, pico-, and nanosecond regime), the laser fluency, the ablation time duration, and the effective liquid medium, with or without the presence of surfactants [1].

The major advantage of both the methods is that it does not include any chemical/reducing and stabilizing agent; therefore, the silver nanoparticles produced by these methods are contamination free and do not need to be purified for further application. However, the major disadvantage is that they consume high energy and costly. Due to these drawbacks, some methods were adopted which are also based on this physical approach but overcome these limitations. These adopted methods are like using ceramic heater which uses less energy and produces continuous heat without any fluctuation and where there is a steep temperature gradient in the vicinity. The second method adopted is thermal decomposition method which produces the silver nanoparticles in solid form. This method works on the principal of complexation between silver and oleate ions and gives silver nanoparticles with 10 nm size. The arch dispersion method was also adopted to overcome the abovementioned limitations and involves the formation of silver nanoparticles in deionized water and does not include the incorporation of any surfactant; it yields silver nanoparticles with less than 10 nm size and hence proves to be a very efficient method [1, 9].

2.2 Chemical methods

These methods are most employed in synthesizing the silver nanoparticles. These methods are based on the reduction of silver ions to the silver atoms which get agglomerated to form the oligomeric clusters which lead to form silver nanoparticles. Various precursors are used in these methods like silver nitrate ($AgNO_3$), silver acetate, and silver chlorate. In these precursors reducing agents like ascorbate, borohydride, and compounds with the hydroxyl and carboxyl group like alcohol, aldehyde, and carbohydrates are incorporated which reduce the silver ion in the precursor and form the silver atom followed by formation of silver nanoparticles. The silver nanoparticles formed are greatly influenced by the nature and properties of reducing agents. The reducing agents are categorized into strong and mild reductants. The strong reductants like borohydrides give large-sized monodispersed nanoparticles, whereas ascorbates and citrates produce small-sized nanoparticles with wide dispersion. Besides this, the morphology (size and shape) of nanoparticles depends upon the type of dispersion medium. The dispersion mediums are a solvent system which acts as the protective or stabilizing agent and is absorbed on the particle surface to prevent agglomeration. Various solvent systems used are mostly polymers like polyvinylpyrrolidone (PVP), polymethylmethacrylate (PMMA), polymethyl acrylic acid (PMAA), and polyethylene glycol (PEG). Polymers are the best candidate as stabilizing agents [10, 11].

2.3 Biological methods

The chemical methods involve a large number of chemical agents like stabilizers (PVP, PMMA, PMAA, and PEG), reducing agents (borohydrides, citrates, and ascorbates) which turns the final product (silver nanoparticles) contaminated. To overcome these limitations, the natural reducing agents are being used nowadays, and this method refers to a biological or green method which is eco-friendly, gives contamination-free product, and consumes less energy. The natural reducing agents like biological microorganisms such as bacteria, fungus, and plant extract are used. The basic principle of this method is that all the natural reducing agents like flavonoids, oils, terpenoids, carbohydrates, enzymes, etc. give an electron to reduce silver ions to silver atoms. This method proves to be a simpler viable alternative to the complex chemical methods to obtain silver nanoparticles. Bacteria are known to be very effective natural reducing agents which give organic and inorganic material, intracellularly and extracellularly. There is a wide range of biological reducing agents available which hence gives a wide choice of precursors for this method (**Figure 2**) [12, 13].

2.4 Other methods

2.4.1 Photochemical method

Photochemical method uses light especially UV light to transform solution of colloidal silver nanoparticles to stable formulation with different sizes and shapes (**Figure 3**). In this method, the precursor source is a silver colloidal solution

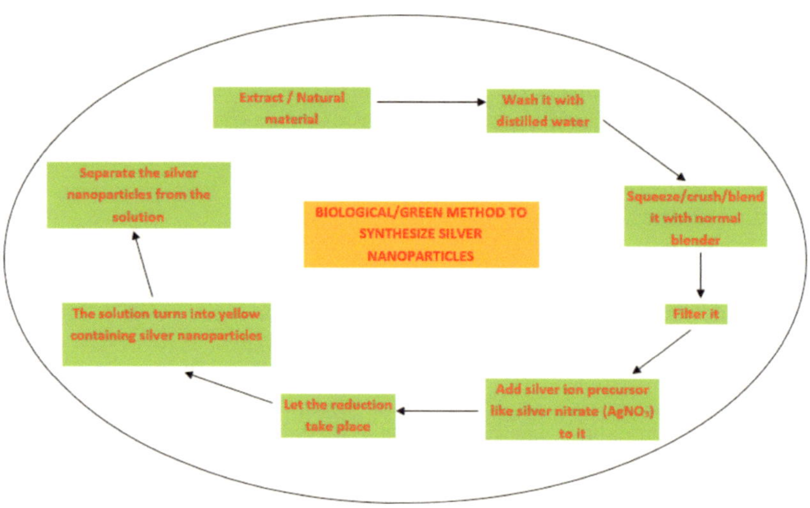

Figure 2.
Overview of the synthesis of silver nanoparticles by the green method.

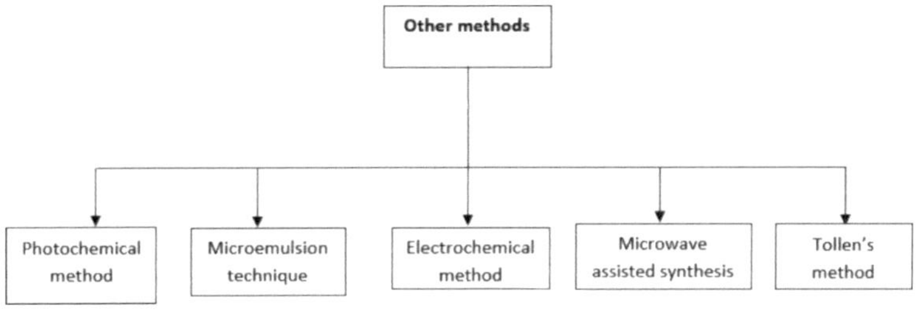

Figure 3.
Other conventional approaches for the synthesis of silver nanoparticles.

(silver nitrate, silver perchlorate, etc.) which gets photochemically reduced to form the silver nanoparticles in the presence of polymer stabilizers such as PVP, PMMA, and PMAA. The growth of the nanoparticles formed by this method can be controlled by choosing the concentration of stabilizers and type of light source [8, 9].

2.4.2 Electrochemical method

In this method, the silver nanoparticles are formed in a special electrochemical cell in which the silver acts as an anode and the platinum acts as a cathode. The external electrical field is applied to the silver anode which in turn forms the silver clusters followed by the formulation of silver nanoparticles that get deposits on the platinum cathode. This process is conducted at the room temperature, and current density can control the size of silver nanoparticles [9].

2.4.3 Microemulsion technique

The silver nanoparticles of controllable and uniform size can be synthesized by this technique. The metal precursor and the reducing agent are firstly separated in the two immiscible liquids; the intensity at the interphase and interphase transporters which are mediated by the quaternary ammonium sulfate affects the rate of interaction between metal precursors and reducing agents. The silver nanoparticle clusters when formed at the interphase get stabilized by the stabilizers at the interphase and then transported to the organic solvents by interphase transporter. The major disadvantages of this method are that the organic solvents which are used are deleterious in nature and that the final product is contaminated in nature and must be separated from the surfactants and organic solvents for further applications which are quite difficult [1].

2.4.4 Microwave-assisted synthesis

In this method, unlike conventional oil bath heating method, microwave heating is used to synthesize the silver nanoparticles. It is a promising method nowadays because microwave heating has a shorter reaction time, reduced energy consumption, and better yield of product which prevent the agglomeration of particles formed. This synthesis involves the carboxymethyl cellulose sodium as a stabilizer. The nanoparticles formed by this have the stability of 2 months without any visual change. Microwave-heated starch is used as a stabilizer which also serves as a template. Polyols like polyethylene glycol and N-vinylpyrrolidine are used as reducing agents as well as stabilizers in which inorganic salt is reduced to form nanoparticles [1].

2.4.5 Tollens method

In this method, the $Ag(NH_3)_2^+$ (Tollens reagent) is reduced by saccharides in the presence of ammonia which yields silver nanoparticle films of size 20–50 nm and silver nanoparticles of different sizes. The pH is usually between the 11.5 and 13.0. pH also influences the particle size as at low pH the size of nanoparticles is comparatively small. The polydispersity of the silver nanoparticles can be achieved by lowering the pH [1].

3. Characterization

3.1 UV-Vis spectroscopy

The absorbance of plasmon is responsible for giving a specific color to the nanoparticles. The electromagnetic radiations and the conduction electron are absorbed by the incident light oscillations and hence produce a specific color. The plasmon sample is diluted with the distilled water generally, and silver nanoparticles show peak near about 400 nm. The lambda max of the plasmon resonance solution is responsible for indicating the size of the formulation (**Figure 4**) [2, 3].

3.2 Fourier transform infrared spectroscopy

In this method, the functional group of the silver nanoparticles is detected. The transmittance goal of silver nanoparticles can be found at 490 nm, and the signaling of OH near 3499 cm [2, 3].

3.3 X-ray diffraction

The XRD depicts the crystalline structure of nanoparticles. When X-rays reflect on the sample (crystal structure), it reflect different diffracted patterns. From these patterns, various physicochemical properties of the sample can be predicted. The X-ray diffraction pattern is matched with the standard/reference pattern of the sample, and from this impurities can be detected easily. There is interplanar spacing in the diffraction pattern which is also called d values; these d values are matched with standard silver values. The average crystalline size of nanoparticles can be calculated using Debye-Scherrer formula:

$$D = \frac{k}{b \cos q} \tag{1}$$

where D is the average crystalline size of the nanoparticles, k is the geometric factor (0.9), λ is the wavelength of X-ray radiation source, and b is the angular full-width at half maximum (FWHM) of the XRD peak at the diffraction angle. From this formula the average size of the silver nanoparticles can be calculated [3, 14].

3.4 Atomic force microscopy

Atomic force microscopy characterizes not only the size shape and sorption but also the dispersion and aggregation of the nanoparticles. AFM helps in the measurement of real-time interactions of nanoparticles with the lipid biological layers, which cannot be achieved by current electron microscopy techniques. No conductive surface or oxide-free surface is required for the measurement in the atomic

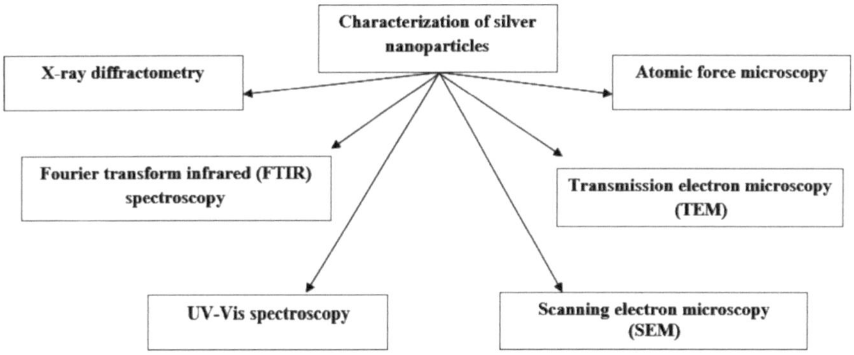

Figure 4.
Various techniques used in the characterization of silver nanoparticles.

force microscopy. In addition to this, the major advantage of AFM is that it does not cause any destruction to the native surface and can measure sub-nanometer scale in aqueous fluids. However, the major drawback is the overestimation of the lateral dimensions of the sample due to the size of the cantilever. The operating mode (no contact or contact) is a very crucial factor in sample analysis [2].

3.5 Scanning electron microscopy

It is a high-resolution technique/microscopy used to detect whole morphology and surface characteristics of the nanoparticles. It is based on the reflection of very high energetic electrons to the probe object. It is a very efficient method to resolve different particle sizes, size distributions, and nanomaterial shapes. The surface morphology of the micro- and nanoscale particles can be easily detected by using SEM. By the histogram obtained particles can be counted either manually or using any software. More specifically for the determination of surface morphology and chemical composition of silver nanoparticles, SEM can be combined with the energy-dispersive X-ray spectroscopy (EDX). The major advantage of this technique is that it can identify the morphology of nanoparticles having size below 10 nm; however, the drawback of this technique is that it is not helpful in determination of the internal structure of the nanoparticles [3].

3.6 Transmission electron microscopy

TEM is a quantitative method for determination of particles, particle size, size distribution, and morphology. In this technique, the resolution is based upon the ratio of distance between the objective lens and specimen and distance between objective lens and image plane. The major advantages of this technique over the SEM are that it has better efficiency of spatial resolution and other analytical measurements can also be done by this technique. However, the major disadvantage of TEM is sample preparation which is a highly crucial step for better imaging and is highly time-consuming; in addition to this, another disadvantage is high-vacuum and very fine and thin sections of sample are required which are quite difficult to maintain and prepare, respectively [2].

4. Applications of silver nanoparticles

Applications of silver nanoparticles can be classified in two major classes, that is, therapeutic and physical applications (**Figure 5**).

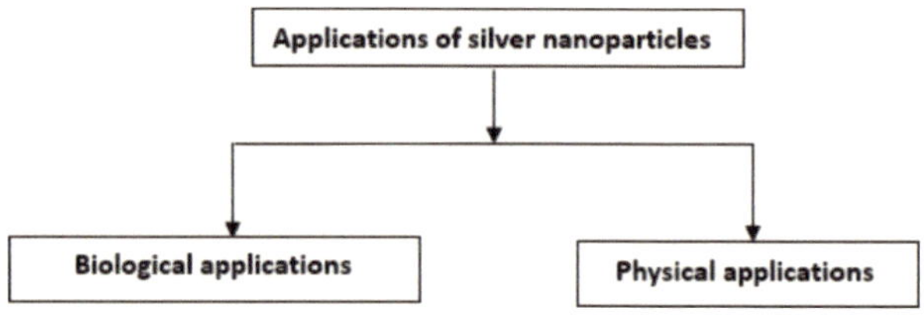

Figure 5.
Applications of silver nanoparticles.

4.1 Biological applications

Silver nanoparticles have various biological applications (**Figure 6**) majorly antimicrobial, anticancer, antioxidant, anti-inflammatory, wound healing, antimalarial, etc. Inbathamiz et al. synthesized silver nanoparticles using aqueous extract of *Morinda pubescens* by reducing silver nitrate and evaluate them in vitro for their antioxidant (using DPPH, ferric thiocyanate, thiobarbituric acid, superoxide anion radical scavenging, H_2O_2, metal chelating, and phosphomolybdenum-like assay) and anticancer potential (by MTT assay on human epithelium cells of liver cancer (HepG2)). They found that silver nanoparticles have high antioxidant capacity as well as cytotoxic activity against HepG2 cell lines [15]. Logeswari et al. synthesized silver nanoparticles using extracts of *Ocimum tenuiflorum*, *Syzygium cumini*, *Solanum trilobatum*, *Centella asiatica*, and *Citrus sinensis* from silver nitrate solution. Prepared silver nanoparticles were evaluated for antimicrobial activity against *Staphylococcus aureus*, *Escherichia coli*, *Klebsiella pneumonia*, and *Pseudomonas aeruginosa* using disk diffusion method. Results revealed that silver nanoparticles synthesized from *Solanum trilobatum* and *Ocimum tenuiflorum* possess the highest antimicrobial activity against *Staphylococcus aureus* (30 mm) and *Escherichia coli* (30 mm), respectively [16].

Most of the urinary tract infections are caused by *Proteus mirabilis*, *Escherichia coli*, *Serratia marcescens*, and *Pseudomonas aeruginosa*. These bacterial pathogens

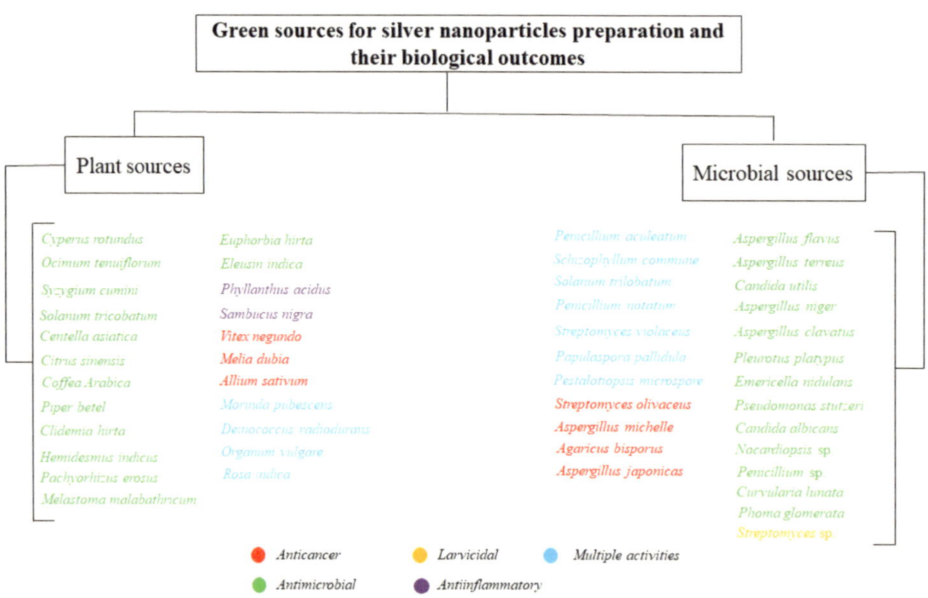

Figure 6.
Natural sources used for preparation of silver nanoparticles and their biological potential.

possess quorum sensing (QS) machinery to coordinate their cells and regulate several virulence factors as well as in biofilm formation. Srinivasan et al. prepared silver nanoparticles using *Piper betle* leaf extract from silver nitrate aqueous solution and evaluate them for anti-QS and antibiofilm potential. Results revealed that prepared silver nanoparticles were able to inhibit QS-mediated virulence factors such as protease, prodigiosin, biofilm formation, and exopolysaccharides as well as hydrophobicity productions in uropathogens. In vivo *Caenorhabditis elegans* assays also revealed their nontoxic and anti-adherence efficacy. Therefore, it was concluded that silver nanoparticles can be an effective alternative toward the conventional antibiotics in controlling QS and biofilm-related uropathogenic infections [17].

Exopolysaccharide of the *Streptomyces violaceus* composed of total carbohydrate (61.4%), ash content (16.1%), and moisture content (1.8%) was efficiently used by Sivasankar et al. for synthesis of silver nanoparticles. Prepared silver nanoparticles were evaluated for antibacterial (against *Pseudomonas aeruginosa*, *Escherichia coli*, *Staphylococcus aureus*, and *Bacillus subtilis*) and antioxidant (using 1,1-diphenyl-2-picrylhydrazyl (DPPH) radical scavenging activity, total antioxidant activity, H_2O_2 scavenging activity, nitric oxide scavenging activity, and ferric reducing power) activities. Results revealed that silver nanoparticles have promising antimicrobial and antioxidant activity [18]. Salama et al. synthesized a series of nanocomposites based on chitosan biguanide-grafted poly(3-hydroxybutyrate) copolymer (ChG-g-PHB) and silver nanoparticles via in situ reduction of silver nitrate in copolymer matrix and evaluated them for antimicrobial activity against *Streptococcus pneumonia*, *Escherichia coli*, *Salmonella typhi*, and *Aspergillus fumigatus*. Results revealed that sample loaded with 3.0% silver nanoparticles has best antimicrobial activity (MIC 0.98–1.95 μg/ml) [19]. Dried roasted *Coffea Arabica* seed extract was used by Dhand et al. for the synthesis of silver nanoparticles from silver nitrate and evaluated for antibacterial potential against *Escherichia coli* and *Staphylococcus aureus*; results confirmed the decrease in bacterial growth with well-defined inhibition zones [20].

Boonkaew et al. developed a burn wound dressing that contains silver nanoparticles to treat infection in a 2-acrylamide-2-methylpropane sulfonic acid sodium salt ($AMPSNa^+$) hydrogel and revealed that hydrogels were nontoxic to normal human dermal fibroblast cells as well as had good action against *Pseudomonas aeruginosa* and methicillin-resistant *Staphylococcus aureus*. They also revealed that 5 mm silver hydrogel was efficient in preventing bacterial colonization of wounds, and results were comparable to the commercially available silver dressings (Acticoat™, PolyMem Silver®) [21]. David et al. did an eco-friendly extracellular biosynthesis of silver nanoparticles using european black elderberry (*Sambucus nigra*) fruit extracts and evaluated them for their in vitro anti-inflammatory activity on HaCaT cells exposed to UVB radiation, in vivo on acute inflammation model, and in humans on psoriasis lesions. Results revealed that silver nanoparticles decrease cytokine production induced by UVB radiation and pre-administration of silver nanoparticles reduces edema and cytokine level in paw tissues after inflammation induction. They also demonstrate the possible use of silver nanoparticles in psoriasis lesions [22]. Silver nanoparticles prepared by chemical reduction from aqueous solution ranged from 10 to 20 nm, and on antibacterial evaluation using Kirby-Bauer method, it was revealed that they have bactericidal activity against *Escherichia coli*, *Pseudomonas aeruginosa*, and *Staphylococcus aureus* [23]. Kathiravan et al. synthesized silver nanoparticles using plant extract of *Melia dubia* and evaluated them against human breast cancer (KB) cell line. Results revealed that prepared silver nanoparticles had remarkable cytotoxicity against KB cell line with high therapeutic index [24]. Latha et al. synthesized silver nanoparticles using leaf extract of *Hemidesmus indicus* and evaluated them for antibacterial activity against the isolated bacteria *Shigella sonnei* using agar bioassay, well diffusion assay, and confocal laser scanning microscopy (CLSM) assay. Results revealed that silver nanoparticles have higher inhibitory activity (34 mm) at 40 μg/ml [25].

Ramar et al. synthesized silver nanoparticles using ethanolic extract of rose (*Rosa indica*) petals and evaluated them for their antibacterial activity against selective human pathogenic microbes and anticancer activity against human colon adenocarcinoma cancer cell line HCT-15. Results revealed that silver nanoparticles were effective against *Escherichia coli*, *Klebsiella pneumonia*, *Streptococcus mutans*, and *Enterococcus faecalis*. The MTT assay, nuclear morphology analysis, mRNA expression of Bcl-2, and Bax and protein expression of caspase 3 as well as caspase 9 indicate the potential anticancer activity [26]. Manikandan et al. prepared silver nanoparticles using aqueous extract of *Phyllanthus acidus* fruits from aqueous silver nitrate solution and investigate their possible role in cytoprotection and anti-inflammation. They find that silver nanoparticles possess potent anti-inflammatory activity by scavenging nitric oxide and superoxide anions [27]. Syafiuddin et al. The silver ions were reduced to silver nanoparticles by using biochemical contents present within *Cyperus rotundus*, *Eleusin indica*, *Melastoma malabathricum*, *Euphorbia hirta*, *Clidemia hirta*, and *Pachyrhizus erosus* extracts. Prepared silver nanoparticles were evaluated for antibacterial capability against *E. coli*, *B. cereus*, and rare bacterium *Chromobacterium haemolyticum*. They found that all silver nanoparticles have antibacterial capability [28]. Pandian et al. synthesized silver nanoparticles using *Allium sativum* extract and evaluated by cytotoxic assays. Surprisingly, prepared silver nanoparticles have enhanced cytotoxic effect and induced many apoptotic cells even with lower concentrations. However, silver nanoparticles are cytotoxic to normal cell line (VERO cells) at higher concentrations, but careful use with lower concentrations can make silver nanoparticles an efficient anticancer agent [29]. Prabhu et al. synthesized silver nanoparticles using leaf extract of *Vitex negundo* as a potential antitumor agent using human colon cancer cell line HCT15. Silver nanoparticles were able to arrest HCT-15 cells at G_0/G_1 and G_2/M phases with a decrease in S phase. Results suggest that silver nanoparticles may exert their antiproliferative effect on colon cancer cell line by suppressing its growth, reducing DNA synthesis, arresting G_0/G_1 phase, and inducing apoptosis [30]. Silver nanoparticles were synthesized by Ramar et al. using unripe fruit extract of *Solanum trilobatum* and evaluated for antibacterial activity against few human pathogenic bacteria (*Streptococcus mutans*, *Enterococcus faecalis*, *Escherichia coli*, and *Klebsiella pneumonia)* as well as anticancer activity against human breast cancer cell line (MCF-7) using MTT, nuclear morphology assay, Western blot, and RT-PCR expression. Results revealed that prepared silver nanoparticles have potential antibacterial and anticancer activities [31]. Silver nanoparticles were evaluated for their effect on growth and health of broiler chickens after infection with *Campylobacter jejuni*, and results revealed that concentration of 50 ppm in drinking water reduces broiler growth and impairs immune functions while having no any antibacterial effect [32].

Sankar et al. prepared silver nanoparticles using the aqueous extract of *Origanum vulgare* by reducing 1 mm silver nitrate solution. They evaluated prepared silver nanoparticles for antibacterial and anticancer efficacy. Silver nanoparticles were found to have an impressive inhibiting effect on human pathogens (*Aeromonas hydrophila*, *Bacillus* spp., *Escherichia coli*, (enteropathogenic—EP), *Klebsiella* spp., *Salmonella* spp., *Salmonella paratyphi*, *Shigella dysenteriae*, and *Shigella sonnei*) as well as a cytotoxic effect against human lung cancer A549 cell line [33]. Sun et al. prepared fabricated silver nanoparticles combined with quercetin, which were stabilized by using a layer of molecules, that is, siRNA, and found that the prepared silver nanoparticles have potential activity against *B. subtilis* [34]. Li et al. synthesized silver nanoparticles by reduction of aqueous silver ion with culture supernatants of *Aspergillus terreus*, and prepared silver nanoparticles showed excellent antimicrobial activity against *Candida albicans*, *Candida krusei*, *Candida parapsilosis*, *Candida tropicalis*, *Aspergillus flavus*, *Aspergillus fumigatus*, *Staphylococcus aureus*, *Pseudomonas aeruginosa*, and *Escherichia coli* [35].

Rajeswari et al. synthesized silver nanoparticles using *Aspergillus* consortium consisting of *Aspergillus niger*, *Aspergillus michelle*, and *Aspergillus japonicus* and evaluated them for anticancer activity against MCF-7 cell line by MTT assay. Results revealed that prepared silver nanoparticles were capable for 100% cell inhibition at 25, 50, and 100 μg concentrations. However, the lowest IC_{50} = 1.47 μg/ml was found for nanoparticles produced from *Aspergillus japonicus* [36]. Sayed et al. synthesized silver nanoparticles using *Aspergillus terreus* cell-free filtrate and evaluated them for antibacterial activity against *Staphylococcus aureus* (MRSA), *Shigella boydii*, *Acinetobacter baumannii*, *Shigella sonnei*, and *Salmonella typhimurium*. They find that prepared silver nanoparticles have potential activity against all the strains [37]. Singh et al. prepared silver nanoparticles using endophytic fungus, *Penicillium* sp., isolated from healthy leaves of *Curcuma longa* and evaluated them against MDR *E. coli* and *S. aureus*. Results revealed that prepared silver nanoparticles have good antibacterial activity with a maximum zone of inhibition of 17 and 16 mm at 80 μL concentration, respectively [38]. Ramalingmam et al. used endophytic fungus *Curvularia lunata* for the extracellular biosynthesis of silver nanoparticles from silver nitrate solution, and prepared silver nanoparticles were tested for antimicrobial potential against *E. coli*, *Pseudomonas aeruginosa*, *Salmonella paratyphi*, *Bacillus subtilis*, *Staphylococcus aureus*, and *Bacillus cereus*. Results revealed that prepared silver nanoparticles have potential antimicrobial activity against all strains [39]. Muhsin et al. synthesized silver nanoparticles using endophytic fungus *Papulaspora pallidula* and evaluated them for antitumor efficacy against human larynx carcinoma cell line (HEp-2). They also investigate them against human pathogenic bacteria (*Escherichia coli*, *Proteus mirabilis*, *Pseudomonas aeruginosa*, *Salmonella typhi*, and *Staphylococcus aureus*) for antibacterial activity using agar well diffusion technique. Results revealed that prepared silver nanoparticles have high inhibition potential against HEp-2 cell line and are effective against all pathogenic bacteria under screening [40].

Arun et al. developed silver nanoparticles using a mushroom fungus *Schizophyllum commune* and evaluated them for their antimicrobial activity against bacterial (*Escherichia coli*, *Bacillus subtilis*, *Pseudomonas fluorescens*, and *Klebsiella pneumonia*) as well as fungal (*Trichophyton simii*, *Trichophyton mentagrophytes*, and *Trichophyton rubrum*) pathogenic strains. They also investigate their anticancer activity using MTT cytotoxicity assay on human epidermoid larynx carcinoma (HEp-2) cell lines. Results revealed that prepared silver nanoparticles have a significant antimicrobial as well as anticancer activity [41]. Barapatre et al. prepared silver nanoparticles by enzymatic reduction of silver nitrate using two lignin-degrading fungi, that is, *Aspergillus flavus* and *Emericella nidulans*, and evaluated them for antibacterial activity against *Escherichia coli* and *Pseudomonas aeruginosa* as well as against *Staphylococcus aureus*. The antibiofilm potential was also tested. They found that prepared silver nanoparticles are effective against tested pathogenic microbes and have the ability to inhibit the biofilm formation by 80–90% [42]. Extracellular synthesis of silver nanoparticles from *Phoma glomerata* was done by Birla et al. and investigated for antibacterial efficacy against *Escherichia coli*, *Staphylococcus aureus*, and *Pseudomonas aeruginosa*. They found that antibiotics showed remarkable sensitivity when used in combination with prepared silver nanoparticles [43].

Subbaiya and Selvam synthesized silver nanoparticles by *Streptomyces olivaceus* and evaluated them for their anticancer potential against non-small cell lung carcinoma cell line (NCI-H460). They found that prepared silver nanoparticles were effective against the cancer cell line [44]. Silver nanoparticles synthesized using aqueous extract of *Agaricus bisporus* fungi were tested for cytotoxic effect on MCF-7 breast cancer cells by El-Sonbaty who found that prepared silver nanoparticles have a dose-dependent cytotoxic effect on MCF-7 breast cancer cells with LD_{50} (50 μg/ml). He also found that silver nanoparticles have a synergistic effect in cancer therapy with gamma radiation [45]. Gade et al. synthesized silver nanoparticles by *Aspergillus niger* isolated from soil and evaluated them for antimicrobial potential. They found that prepared silver nanoparticles have remarkable antibacterial activity against *Staphylococcus aureus* and *Escherichia coli*, respectively [46]. Endophytic fungal species, *Penicillium* species from *Glycosmis mauritiana*, was used for the synthesis of silver nanoparticles by Govindappa et al and evaluated for their biological potential. They found that prepared silver nanoparticles have anti-inflammatory, xanthine oxidase, and lipoxygenase and tyrosine kinase inhibitory activity. Furthermore, prepared silver nanoparticles strongly inhibit bacterial species like *E. coli* and *P. aeruginosa* [47].

Rajam et al. prepared silver nanoparticles using fungus *Emericella nidulans* EV4 and investigated their potential against *Pseudomonas aeruginosa* NCIM 5029. They found that prepared silver nanoparticles showed remarkable control over the growth of *Pseudomonas aeruginosa* NCIM 5029 [48]. Kulkarni et al. synthesized silver nanoparticles using *Deinococcus radiodurans* and found that prepared silver nanoparticles have remarkable antimicrobial, anticancer, and anti-biofouling activity [49]. Netala et al. prepared silver nanoparticles by using the aqueous culture of filtrate from *Pestalotiopsis microspora* and evaluated them for antioxidant and anticancer potential. Prepared silver nanoparticles showed remarkable radical scavenging activity against DPPH and H_2O_2 radicals with IC_{50} values of 76.95 ± 2.96 and 94.95 ± 2.18 μg/ml as well as significant cytotoxic effect against SKOV3 (human ovarian carcinoma, IC_{50} = 16.24 ± 2.48 μg/ml), B16F10 (mouse melanoma, IC_{50} = 26.43 ± 3.41 μg/ml), PC3 (human prostate carcinoma, IC_{50} = 27.71 ± 2.89 μg/ml), and A549 (human lung adenocarcinoma, IC_{50} = 39.83 ± 3.74 μg/ml) cells, respectively [50]. Silver nanoparticles prepared by using the cell-free extract of *Saccharomyces boulardii* were tested for anticancer activity against breast cancer cell lines (MCF-7 cells) by Kaler A. et al. who found that silver nanoparticles showed very high activity on MCF-7 cells, showing 80% inhibition [51]. Durairaj et al. synthesized silver nanoparticles using *Penicillium notatum* and evaluated them for their antibacterial and larvicidal potential in mosquitoes. They found that silver nanoparticles have significant mortality rate against second and third instar larvae of *Culex quinquefasciatus* after 24 h exposure and were effective against *Staphylococcus aureus*, *Escherichia coli*, *Salmonella Shigella*, and *Salmonella typhimurium*, respectively [52]. Silver nanoparticles prepared by using culture supernatants of *Aspergillus terreus* were evaluated for their antimicrobial properties by Li et al. and found to have broad-spectrum antimicrobial activity against *A. terreus* against *P. aeruginosa*, *S. aureus*, *E. coli*, *C. albicans*, *C. krusei*, *C. glabrata*, *C. tropicalis*, *A. fumigatus*, and *A. flavus*, respectively [35]. Ma et al. prepared silver nanoparticles using cell-free filtrate of the fungus strain *Penicillium aculeatum* Su1 as reducing agent and found that prepared nanoparticles exhibit higher antimicrobial activity against *E. coli*, *P. aeruginosa*, *S. aureus*, *B. subtilis*, and *C. albicans* as well as have higher biocompatibility toward human bronchial epithelial (HBE) cells with high cytotoxicity in dose-dependent manner (IC_{50} = 48.73 μg/ml) toward A549 cells [53].

Silver nanoparticles were prepared using culture supernatant of *Nocardiopsis* sp. MBRC-1 and evaluated for antimicrobial and anticancer activity by Manivasagan et al. Results revealed that silver nanoparticles have strong antimicrobial activity against bacteria (*Escherichia coli*, *Pseudomonas aeruginosa*, *Bacillus subtilis*, *Enterococcus hirae*, *Shigella flexneri*, and *Staphylococcus aureus*) and fungi (*A. brasiliensis*, *A. fumigatus*,

Aspergillus niger, and *Candida albicans*) and are effective against human cervical cancer cell line (HeLa) in dose-dependent manner with IC_{50} value of 200 μg/ml, respectively [54]. Silver nanoparticles were synthesized by Rahimi et al. using biomass obtained from the culture of *Candida albicans* and evaluated for antibacterial properties. Results revealed that prepared silver nanoparticles were effective against *Escherichia coli* and *Staphylococcus aureus* [55]. Rajora et al. used textile soil-isolated bacterium *Pseudomonas stutzeri* to synthesize silver nanoparticles and evaluated them for their antimicrobial and cytotoxicity properties. Results revealed that prepared silver nanoparticles have strong antibacterial activity against multidrug-resistant (MDR) *Escherichia coli* and *Klebsiella pneumonia* and do not have any cytotoxic effects on human epithelial cells [56]. Shanmugasundaram et al. isolated an actinobacterium, *Streptomyces* sp. M25, and used its biomass for the synthesis of silver nanoparticles. Prepared silver nanoparticles when evaluated for larvicidal activity were found to have significant activity against malarial vector, *Anopheles subpictus* (LC_{50} = 51.34 mg/l) and filarial vector *Culex quinquefasciatus* (LC_{50} = 60.23 mg/l), respectively [57].

Kalaivani et al. prepared silver nanoparticles using *Lactobacillus acidophilus* and white rot fungus (*Pleurotus platypus*) and found that prepared silver nanoparticles were effective when evaluated for antibacterial potential against pathogenic bacterial strains such as *Bacillus subtilis*, *Escherichia coli*, and *Staphylococcus aureus*, respectively [58]. Silver nanoparticles prepared by using *Aspergillus clavatus* (AzS-275), an endophytic fungus isolated from sterilized stem tissue of *Azadirachta indica* A. Juss, were evaluated for antimicrobial potential and found effective against *Candia albicans* (MIC = 9.7 μg/ml, inhibition zone = 16 mm), *Pseudomonas fluorescens* (inhibition zone = 14 mm), and *Escherichia coli* (inhibition zone = 10 mm), respectively [59]. Waghmare et al. prepared silver nanoparticles using *Candida utilis* NCIM 3469 and evaluated them for antibacterial potential. Results revealed that prepared silver nanoparticles were effective against pathogenic organisms such as *Pseudomonas aeruginosa* (inhibition zone = 13 ± 1.2 mm), *Staphylococcus aureus* (inhibition zone = 8 ± 0.8 mm), and *Escherichia coli* (inhibition zone = 10 ± 1.0 mm), respectively [60].

4.2 Physical applications

4.2.1 Fabrication of antennas

Alshehri et al. have prepared two samples: the first was fabricated from the nano-metallic silver, and the second consists of micrometer-sized grains. Both types were prepared using thick-film fabrication process. The material involved in sample preparation was fine metal powder, an inorganic binder-like metal oxide, and an organic vehicle that evaporates during the initial drying stages. Both the samples were characterized for the electrical performances. They found that in the lower-frequency range, both types of conductors (samples) behave similarly with electrical loss but increase approximately linearly with increased frequency range (from 0.1 dB/mm/GHz up to 80 GHz), but above 80 GHz frequency, the silver nanoparticle-fabricated sample showed lower electrical loss, and this behavior continues up to the above whole frequency range. The lower level of the loss from the silver nanoparticle conductors and the overall trend in loss per wavelength do not depend significantly on frequency. Therefore, it has been concluded that the silver nanoparticle-fabricated conductors show a less electrical loss at high-frequency range which in turn attributed to lower surface roughness found in the nanoparticles due to better packing and may open opportunities for low-temperature fabrication of antennas and for sub-THz metamaterials with improved performance [61].

4.2.2 In electronically conductive adhesives

Silver nanoparticles can be used as a silver paste in the electrodes because of their high conductivity. They have also been used as conductive fillers in electronically conductive adhesives (ECAs). Chen et al. have synthesized the silver nanoparticles by reducing the silver nitrate with ethanol in the presence of polyvinylpyrrolidone (PVP). Various reaction conditions have been studied such as PVP concentration, reaction time, and reaction temperature. In this method, PVP prevents the aggregation; in addition to this, the PVP increases the rate of spontaneous nucleation and decreases the mean size of silver nanoparticles. The ethanol used in this has been employed as a reducing agent or diluent to adjust the viscosity of the ECAs. The resulting silver nanoparticles obtained with chemical reduction method had very fine dispersion and narrow size distribution. The ECAs had the silver nanoparticles re-dispersed in the ethanol. The absorption peak was determined at 410 nm which was a clear signature of the quantum size effect occurring in the absorption property of silver nanoparticles. It has also been concluded that the particle size of nanoparticles has been decreased with increasing concentration of silver nitrate and with increasing reaction temperature, but with increasing reaction time, the size of nanoparticles has been increased [62]. Yang et al. have prepared silver nanoparticles, silver nanorods, and epoxy resins containing high-performance electrically conductive adhesives (ECAs) using a novel preparation method. The prepared nanoparticles and nanorods were dispersed well, and there was no agglomerate in the matrix. The volume electrical resistivity tests showed the volume electrical resistivity of the ECA was closely related with the various sintering temperatures and time and time and the ECA could achieve the volume electrical resistivity of (3–4) $\times$ 9 10^{-5} Ω after sintering at 160°C for 20 min. They found that the prepared ECA was able to achieve low-temperature sintering and possessed excellent electrical, thermal, and mechanical properties [63]. This offers the possibility to effectively use these synthesized nanoparticles for improving the conductivity of ECAs.

4.2.3 Ink-jet printing

The silver nanoparticles can be used in ink-jet printing. Wu and Hsu have synthesized the silver nanoparticles by chemical reduction from the silver nitrate using triethylamine as reducing and protecting agent. After that the nanoparticles have been sintered using the process involved cleaning it with acetone and deionized water to remove the particles and organic contaminants on the surface; after cleaning the film, it was treated with ozone by UVO-100 UV ozone for 30 min. The silver nanoparticle suspensions were spin coated (500 rpm, 15 s) on the polyimide substrate and dried at room temperature in order to remove the solvent. The resulting silver nanoparticles on the polyimide substrate were heated from 100 to 200°C and held at 200°C for 1 h in order to convert to silver films. The polyimide substrate was then naturally cooled at room temperature in the glass dish. The above synthesized silver nanoparticles were sintered at different temperatures, and it was found that the resistivity of the silver film sintered at 150°C for 1 h was close to the resistivity of bulk silver. Based on the above data, the synthesized nanoparticles had the low sintering temperature; hence, the silver nanoparticle suspensions could be used to fabricate the flexible electronics by ink-jet printing [64].

4.2.4 Fillers

The micro-sized silver particle fillers appear as the full-density silver flakes, and the silver nanoparticles fillers appear to be the highly porous agglomerates (similar to open-cell foams). Ye measured/analyzed the distribution of different sized particles using TEM. The electrical resistivity was also measured which was compared with the different levels of filler loading. The silver nanoparticles were prepared using the nano-sized spheres of size approximately 50–150 nm in diameter, micro-sized particles with a diameter of 5–8 μm, and flakes of silver of 10 μm in length. By TEM studies of the distribution of silver particles in micro-sized particle sea, it was concluded that it is difficult to find the cross-linkage of particles and there are fewer chances of different contact and contact area, and by the resistivity measurements, it has been revealed that the conductivity of micro-sized silver particle-filled adhesive is dominated by constriction resistance, the silver nanoparticle-containing adhesive is controlled by tunneling and even thermionic emission, and hence the respective nanoparticles are used to increase the electrical conductivity [65].

4.2.5 Water treatment

Dankovich prepared silver nanoparticles in a paper using microwave irradiation. Antibacterial activity and silver release from the silver nanoparticle sheets were assessed for model *Escherichia coli* and *Enterococcus faecalis* bacteria in deionized water and in suspensions that also contained with various influent solution chemistries, that is, with natural organic matter, salts, and proteins. The paper sheets containing silver nanoparticles were effective in inactivating the test bacteria as they passed through the paper. The resultant silver nanoparticle paper is just as effective for inactivating bacteria during percolation through the sheet; the silver nanoparticle papers effectively purify water contaminated with bacteria. Hence, in conclusion, the paper incorporated with silver nanoparticles by microwave has been used for the purification of contaminated water [66]. Park et al. developed micrometer-sized silica hybrid composite decorated with silver nanoparticles, that is, AgNP-SiO_2 (to prevent the inherent aggregation of silver nanoparticles and easy recovery from environmental media after utilization), and evaluated them for antiviral activity using bacteriophage MS2 and murine norovirus (MNV) models. Results revealed their potential, and it was concluded that developed silver nanoparticles (AgNP-SiO_2) can be efficiently used in disinfection processes for inactivation of various waterborne viruses [67]. Abu-Elala et al. investigated the effect of chitosan-silver nanocomposite on fish crustacean parasite, *Lernaea cyprinacea*, disease found in goldfish (*Carassius auratus*) aquaria during the spring season. Their results proposed that chitosan silver nanocomposite is efficient in parasitic control in ornamental glass aquaria [68].

4.2.6 Solar cell optimization

Plasmonic effects in thin film silicon solar cell are an emerging technology and area of rigorous research for the researchers from the past couple of years. It has promising application in solar cell fabrication industries where it uses nanoscale properties of silver nanoparticles incorporated in the interface between the metal and dielectric contacts that enhance the light-trapping properties of thin film silicon solar cells by increase absorbance capacity and generation of hot electrons that enhance the photocurrents in the solar cell. Sangno et al. had taken two different thicknesses of the silver thin film (made of silver nanoparticles) of 5.9 and 7.8 nm in 2×10^{-4} (Torr) and 2.5×10^{-4} (Torr) pressure environment for investigation purpose. Samples were annealed at different temperature ranges for a definite time period under vacuum condition of 4.5×10^{-6} Torr. They found that reflectance reduces 13–11% due to plasmonic effect and enhancement in the conversion efficiency of the solar cell [69].

4.2.7 Biosensor fabrication

Li et al. fabricated nanoenzymatic glucose biosensors by depositing silver nanoparticles using in situ chemical reduction method on TiO_2 nanotubes which were synthesized by the anodic oxidation process. The structure, morphology, and mechanical behaviors of the electrode were examined by scanning electron microscopy and nanoindentation. It was found that silver nanoparticles remained both inside and outside of TiO_2 nanotubes whose length and diameter were about 1.2 μm and 120 nm. The composition was constructed as an electrode of a non-enzymatic biosensor for glucose oxidation. The electrocatalytic properties of the prepared electrodes for glucose oxidation were investigated by cyclic voltammetry (CVs) and differential pulse voltammetry (DPV). When compared with bare TiO_2 and silver-fresh TiO_2 nanotube, Ag-TiO_2/(500°C) nanotube exhibited the best electrochemical properties from cyclic voltammetry (CVs) results. In addition, the nonenzymatic glucose sensors exhibited excellent selectivity, stability, and repeatability. Nanoenzymatic glucose biosensors have potential application in catalysis and sensor areas [70]. Ruth et al. has synthesized the oligonucleotide-silver nanoparticle (OSN) conjugates and revealed their use with magnetic beads as a biosensor for *Escherichia coli* detection under the magnetic field condition. The biosensor developed was able to detect the presence of DNA target which was isolated from the three isolation methods, and it has been found that best detection signals were achieved by the isolation method in which it could detect the presence of DNA target up to 1.3 ng/μl [71]. Mahmudin et al. synthesize the silver nanoparticles by chemical reduction method. TEM images showed that morphology of silver nanoparticles had spherical geometry and had dispersive particle distribution. They conclude that this type of dispersibility of nanoparticles such as this could potentially be used as an active ingredient of SPR biosensor [72]. Sistani et al. have developed the enzymatic biosensor for selective detection of penicillin by using silver nanoparticles, and sensor configuration showed the linear dynamic range for output response vs. logarithmic concentration of a salt solution of penicillin G [73].

4.2.8 Protein sensing

Tung N.H reported that silver nanoparticles labeling could be used in protein sensing studies by liquid electrode plasma-atomic emission spectrometry (LEP-AES). This technique is suitable for on-site portable analysis because plasma gas and the high-power source are not required. Proposed detection method could have a wide variety of promising applications in metal nanoparticle-labeled biomolecule detection [74].

4.2.9 Hospitals

Duran et al. prepared silver nanoparticles by using *Fusarium oxysporum* and studied their antimicrobial effect when incorporated in cotton fabrics against *Staphylococcus aureus*. They found that fabric incorporated with silver nanoparticles have significant antibacterial activity. They proposed that clothes with silver nanoparticles are sterile and can be useful in hospitals to prevent or to minimize infection with pathogenic bacteria such as *Staphylococcus aureus* [75].

4.2.10 Analytical

Lipids are the major components of cell membrane and abnormal cellular metabolism-induced lipid changes. Hua et al. investigate silver nanoparticle-induced lipid changes on the surface of macrophage cells using time-of-flight secondary ion mass spectrometry (ToF-SIMS). By using this technique, one can understand the mechanism of cell-nanoparticle interactions at the molecular level and characterize the changes in lipids on the single cell surface [76]. Citrate- and polyethyleneimine-coated silver nanoparticles can be used to understand how the type of capping agents and surface charge affects their colloidal stability, dissolution, and ecotoxicity in the absence/presence of Pony Lake fulvic acid (PLFA). On the basis of this, Jung et al. demonstrate that the differences in colloidal stability, ecotoxicity, and dissolution may be attributed to different capping agents, surface charge, and natural organic matter concentration as well as to the formation of dissolved silver complexes with natural organic matter [77].

4.2.11 Agricultural and marine

Silver nanoparticles synthesized by Guilger et al. using fungus *Trichoderma harzianum* were evaluated for cytotoxicity and genotoxicity against fungus *Sclerotinia sclerotiorum* which is responsible for the agricultural disease white mold and found that nanoparticles showed potential against *Sclerotinia sclerotiorum*, inhibiting sclerotium germination and mycelial growth. The study suggests that silver nanoparticles can be a new alternative in white mold control [78]. Babu et al. have synthesized silver nanoparticles in vitro using marine bacteria *Shewanella algae bangaramma* and found that the synthesized nanoparticles have both larvicidal and bactericidal activities and no mortality in control; in addition to this, the maximum values of LC_{50} and LC_{90} with 95% confidential limit [4.529 mg/ml (2.478–5.911), 9.580 mg/ml (7.528–14.541)] were observed with third instar larvae of *Lepidiota mansueta* (Burmeister). It was found that the mortality of larvae was significantly increased in all the concentrations ($P < 0.0001$) in all the exposed groups. The bactericidal activities of the silver nanoparticles were determined against some of the bacterial species which followed the following order: *Vibrio cholera* < *Roseobacter* spp. < *Alteromanas* spp. It has been concluded that the synthesized silver nanoparticles had effective larvicidal and antifouling activities and can be effectively used in the agricultural and marine pest control [79].

4.2.12 Miscellaneous

Chen prepared silver nanoparticles from filamentous fungus *Phoma* sp3.2883 via adsorption and accumulation as well as proposed that fungus *Phoma* sp3.2883 is a potential biosorbent that can be used for the production of silver nanoparticles and would be useful in waste detoxification and in silver recovery programs [80]. Du et al. synthesized silver nanoparticles under light radiation using cell filtrate of *Penicillium oxalicum* 1–208. The prepared silver nanoparticles were used as a catalyst and exhibit excellent catalytic activity for reduction of methylene blue in the presence of $NaBH_4$ at ambient temperature [81]. Otari et al. synthesized silver nanoparticles using culture supernatant of phenol degraded broth (prepared by using an actinobacterium *Rhodococcus* NCIM 2891) and investigate their catalytic potential. They found that prepared silver nanoparticles have excellent catalytic activity in the reduction of 4-nitrophenol to 4-aminophenol by $NaBH_4$ [82]. Zaheer synthesized silver nanoparticles using an aqueous extract of date palm fruit pericarp and evaluated them for antimicrobial activity against multiple drug-resistant *Staphylococcus aureus*, *Escherichia coli*, and *Candida albicans*. Results revealed that inhibition was concentration dependant. It was also concluded that silver nanoparticles have good catalytic activity toward the catalytic and photocatalytic degradation of 4-nitrophenol [83]. Soni et al. prepared silver nanoparticles using soil fungus *Aspergillus niger* 2587 and evaluated them against larvae and pupae of *Anopheles stephensi*, *Culex quinquefasciatus*, and *Aedes aegypti*. Results revealed that larvae of *Culex quinquefasciatus* were most susceptible and showed 100% mortality after 1 h of exposure. This suggests the possible application of silver nanoparticles in mosquito control [84]. Larvicidal potential of silver nanoparticles synthesized by using filamentous fungus *Cochliobolus lunatus* was determined against vectors *Aedes aegypti* and *Anopheles stephensi* by Salunkhe et al. They found that silver nanoparticles have efficacy against the second, third, and fourth instar larvae of *A. stephensi* (LC_{50} 1.17, 1.30, and 1.41; LC_{90} 2.99, 3.13, and 3.29 ppm) and against *A. aegypti* (LC_{50} 1.29, 1.48, and 1.58; LC_{90} 3.08, 3.33, and 3.41 ppm). They also proposed that possible larvicidal activity may be due to the penetration of nanoparticles through the membrane [85]. Sanago et al. investigated silver nanoparticles synthesized by using filamentous fungus *Penicillium verrucosum*, for larvicidal activities against the filarial vector *Culex quinquefasciatus*. They find that synthesized silver nanoparticles were effective against the first, second, third, and fourth instar larvae and pupae of *Culex quinquefasciatus* with LC_{50} (LC_{90}) values of 4.91 (8.13), 5.16 (8.44), 5.95 (7.76), and 7.83 (12.63) at 25 ppm as well as 5.24 (8.66), 5.56 (8.85), 6.20 (10.01), 7.04 (10.92) and 7.33 (11.59) at 50 ppm in larval instars and pupae [69]. Silver nanoparticles were prepared by Suresh et al. using an aqueous extract of *Delphinium denudatum* and evaluated for their larvicidal (against second instar larvae of the dengue vector *Aedes aegypti*) potential. They found that prepared silver nanoparticles have potent larvicidal activity with LC_{50} value of 9.6 ppm [86].

5. Conclusion

It is revealed that silver nanoparticles have potential applications in therapeutics as well as in other physical fields. In therapeutics, researchers are seemed to be more focused on anticancer and antimicrobial evaluations. Green synthesis makes them eco-friendly and nonhazardous. Applications of silver nanoparticles are not limited to therapeutics only, they are equally covering physical fields too such as biosensors and antenna fabrication, conductive adhesives, in ink-jet printing, water treatment, solar cell optimization, protein sensing, etc. Rigorous research has been carried out and continued on this nanostructure. Therefore, the silver nanoparticle has the ability to be a lead nanoparticle of the future due to its wide variety of applications.

Author details

Atamjit Singh[1*] and Kirandeep Kaur[2]

1 Laureate Institute of Pharmacy, Kathog, Himachal Pradesh, India

2 Shaheed Bhagat Singh Polytechnic and Pharmacy College, Patti, Punjab, India

*Address all correspondence to: atampanesar@gmail.com

References

[1] Iravani S, Korbekandi H, Mirmohammadi SV, Zolfaghari B. Synthesis of silver nanoparticles: chemical, physical and biological methods. Research in Pharmaceutical Sciences. 2014;**9**(6):385-406

[2] Zhang X, Liu Z, Shen W, Gurunathan S. Silver nanoparticles: Synthesis, characterization, properties, applications and therapeutic approaches. International Journal of Molecular Sciences. 2016;**17**:1-34

[3] Tran QH, Nguyen VQ, Le A. Silver nanoparticles: Synthesis, properties, toxicology, applications and perspectives. Advances in Natural Sciences: Nanoscience and Nanotechnology. 2013;**4**:1-21

[4] Rauwel P, Rauwel E, Ferdov S, Singh MP. Silver nanoparticles: Synthesis, properties and applications. Advances in Materials Science and Engineering. 2015;**2015**:1-2

[5] Wie L, Lu J, Xu H, Patel A, Chen Z, Chen G. Silver nanoparticles: Synthesis, properties and therapeutic applications. Drug Discovery Today. 2015;**20**(5):595-601

[6] Bartlomiejczyk T, Lankoff A, Kruszewski M, Szumiel I. Silver nanoparticles-allies or adversaries. Annals of Agriculture and Environmental Medicine. 2013;**20**(1):48-54

[7] Alaqad K, Saleh TA. Gold and silver nanoparticles: Synthesis methods, characterization routes and applications toward drugs. Journal of Environmental and Analytical Toxicology. 2016;**6**(4):1-10

[8] Haider A, Kang IK. Preparation of silver nanoparticles and their industrial and biomedical applications: A comprehensive review. Advances in Materials Science and Engineering. 2015;**2015**:1-16

[9] Zhang S, Tang Y, Vlahovic B. A review on preparation and application of silver nanofibres. Nanoscale Research Letters. 2016;**11**(80):1-8

[10] Khouloud MM, Abou E, Eftaiha A, Abdulrhman A, Reda A. Synthesis and applications of silver nanoparticles. Arabian Journal of Chemistry. 2010;**3**:135-140

[11] Natsuki J, Natsuki T, Hashimoto Y. A review of silver nanoparticles: Synthesis methods, properties and applications. International Journal of Material Science and Applications. 2015;**4**(5):325-332

[12] Ahmed S, Ahmed M, Swami BL, Ikram S. A review on plant extract mediated synthesis of silver nanoparticles for antimicrobial applications: A green expertise. Journal of Advanced Research. 2016;**7**:17-28

[13] Srikar SK, Giri DD, Pal DB, Mishra PK, Upadhyay SN. Green synthesis of silver nanoparticles: A review. Green and Sustainable Chemistry. 2016;**6**:34-56

[14] Jyoti K, Baunthiyal M, Singh A. Characterization of silver nanoparticles synthesized using *Urtica dioica* Linn. Leaves and their synergistic effects with antibiotics. Journal of Radiation Research and Applied Sciences. 2016;**9**(2016):217-227

[15] Inbathamiz L, Ponnu TK, Mary EJ. In vitro evaluation of antioxidant and anticancer potential of morinda pubescens synthesized nanoparticles. Journal of Pharmacy Research. 2013;**6**:32-38

[16] Logeswari P, Silambarasan S, Abraham J. Synthesis of silver nanoparticles using plant extract and analysis of their antimicrobial property.

Journal of Saudi Chemistry Society. 2015;**19**:311-317

[17] Srinivasan R, Vigneshwari L, Rajavel T, Durgadevi R, Kannapan A, Balamurugan K, et al. Biogenic synthesis of silver nanoparticles using *Piper betle* aqueous extract and evaluation of its anti-quorum sensing and antibiofilm potential against uropathogens with cytotoxic effects: an in vitro and in vivo approach. Environmental Science and Pollution Research. 2018;**25**:10538-10554. DOI: 10.1007/s11356-017-1049-0

[18] Sivasankar P, Seedevi P, Poongodi S, Sivakumar M, Murugan T, Sivakumar L, et al. Characterization. Antimicrobial and antioxidant property of exopolysaccharide mediated silver nanoparticles synthesized by Streptomyces violaceus MM72. Carbohydrate Polymers. 2018;**181**:752-759. DOI: 10.1016/j.carbpol.2017.11.082

[19] Salama HE, Aziz MSA, Riad G. Thermal properties, crystallization and antimicrobial activity of chitosan biguanidine grafted poly (3-hydroxybutyrate) containing silver nanoparticles. Biomacromolecules. 2017. DOI: 10.1016/j.ijbiomac.2017.12.153

[20] Dhand V, Soumya L, Bharadwaj S, Chakra S, Bhatt D, Sreedhar B. Green synthesis of silver nanoparticles using coffea Arabica seed extract and its anti-bacterial activity. Material Science and Engineering C. 2016;**58**:36-43

[21] Boonkaew B, Barber PM, Rengipipat S, Supaphol P, Kemph M, He J, et al. Development and characterization of a novel, antimicrobial, sterile hydrogel dressing for burn wounds: Single step production with gamma irradiation creates silver nanoparticles and radical polymerization. Pharmaceutical Nanotechnology. 2014;**103**:3244-3253. DOI: 10.1002/jps. 24095

[22] David L, Moldovan B, Vulcu A, Olenic L, Schrepler MP, Fodor EF, et al. Green synthesis, characterization and anti-inflammatory activity of silver nanoparticles using European black elderberry fruit extract. Colloid and Surfaces B: Biointerfaces. 2014;**122**:767-777. DOI: 10.1016/j.colsurfb.2014.08.018

[23] Guzman M, Dille J, Godet S. Synthesis and antibacterial of silver nanoparticles against gram-positive and gram-negative bacteria. Nanomedicine: Nanotechnology, Biology and Medicine. 2012;**8**:37-45

[24] Kathiravan V, Ravi S, Ashokkumar S. Synthesis of silver nanoparticles from melia dubia leaf extract and their in vitro anticancer activity. Spectrochimica Acta Part A: Molecular and Biomolecular Spectroscopy. 2014;**130**:116-121

[25] Latha M, Sumathi M, Manikandan R, Arumugam A, Prabhu NM. Biocatalytic and bactericidal interaction visualization of green synthesized silver nanoparticles using hemidesmus indicus. Microbial Pathogenesis. 2015;**82**:43-49. DOI: 10.1016/j.micpath .2015.03.008

[26] Ramar M, Manikandan B, Raman T, Arunagirinathan K, Prabhu NM, Babu MJ, et al. Biosynthesis of silver nanoparticles using ethanolic petals extract of *Rosa indica* and characterization of its antibacterial, anticancer and anti-inflammatory activities. Spectrochimica Acta Part A: Molecular and Biomolecular Spectroscopy. 2015;**138**:120-129. DOI: 10.1016/j. saa.2014.10.043

[27] Manikandan R, Beulaja M, Thiagarajan R, Palanisamy S, Goutham G, Koodalingam A, et al. Biosynthesis of silver nanoparticles using aqueous extract of *Phyllanthus acidus* L.fruits and characterization of its anti-inflammatory effect against H_2O_2-exposed rat peritoneal macrophages. Process Biochemistry. 2017;**55**:72-81. DOI: 10.1016/j. procbio.2017.01.023

[28] Syafiuddin A, Salmiati, Hadibarata T, ABH K, Salim MR. Novel weed-extracted silver nanoparticles and their antibacterial appraisal against a rare bacterium from river and sewage treatment plan. Nanomaterials. 2008;**8**(9). DOI: 10.3390/nano8010009

[29] Pandian AMK, Karthikeyan C, Rajasimman M, Dinesh MG. Synthesis of silver nanoparticles and its application. Ecotoxicology and Environmental Safety. 2015;**121**:211-217. DOI: 10.1016/j.ecoenv.2015.03.039i

[30] Prabhu D, Arulvasu C, Babu G, Manikandan R, Srinavasan P. Biologically synthesized green silver nanoparticles from leaf extract of *Vitex negundo* L. Induce growth-inhibitory effect on human colon cancer cell line HCT15. Process Biochemistry. 2013;**48**:317-324

[31] Ramar M, Manikandan B, Marimuthu PN, Raman T, Mahalingam A, Subramanian P, et al. Synthesis of silver nanoparticles using *Solanum trilobatum* fruits extract and its antibacterial, cytotoxic activity against human breast cancer cell line MCF 7. Spectrochimica Act. 2015;**140**:223-228. DOI: 10.1016/j.saa.2014.12.060

[32] Vadalasetty KP, Lauridsen C, Engberg RM, Vadalasetty R, Kutwin M, Chwalibog A, et al. Influence of silver nanoparticles on growth and health of broiler chickens after infection with *Campylobacter jejuni*. BMC Veterinary Research. 2018;**14**(1):1-11

[33] Sankar R, Karthik A, Prabu A, Karthik S, Shivashangari KS, Ravikumar V. Origanum vulgare mediated biosynthesis of silver nanoparticles for its antibacterial and anticancer activity. Colloids and Surfaces B: Biointerfaces. 2013;**108**:80-84

[34] Sun D, Zhang W, Zhao Z, Li N, Mou Z, Yang E, et al. Silver nanoparticles-quercetin conjugation to siRNA against drug-resistant *Bacillus subtilis* for effective gene silencing: in vitro and in vivo. Material Science and Engineering C. 2016;**63**:522-534

[35] Li G, Qian Y, He D, Guan B, Gao S, Cui Y, et al. Fungus-mediated green synthesis of silver nanoparticles using *Aspergillus terreus*. International Journal of Molecular Sciences. 2012;**13**:466-476

[36] Rajeswari P, Samuel P, Vijayakumar J, Selvarathinam T, Sudarmani DNP, Amirtharaj K, et al. Green synthesis of silver nanoparticles by aspergillus consortium and evaluating its anticancer activity against human breast adenocarcinoma cell line (MCF7). Pharmaceutical and Biological Evaluations. 2017;**4**(1):28-36

[37] Sayed SRM, Bahkali AH, Bakri MM, Hirad AH, Elgorban AM, Metwally MA. Antibacterial activity of biogenic silver nanoparticles produced by *Aspergillus terreus*. International Journal of Pharmacology. 2015;**11**(7):858-863

[38] Singh D, Rathod V, Ninganagouda S, Hiremath J, Singh AK, Mathew J. Optimization and characterization of silver nanoparticle by endophytic *Fungi penicillium* sp. isolated from *Curcuma longa* (turmeric) and application studies against MDR *E. coli* and *S. aureus*. Bioinorganic Chemistry and Applications. 2014. DOI: 10.1155/2014/408021

[39] Ramalingmam P, Muthukrishnan S, Thangaraj P. Biosynthesis of silver nanoparticles using an endophytic fungus, curvularialunata and its antimicrobial potential. Journal of Nanoscience and Nanoengineering. 2015;**1**(4):241-247

[40] Muhsin T, Hachim A. Antitomor and antibacterial efficacy of mycofabricated silver nanoparticles by endophytic fungus *Papulaspora pailidula*. American Journal of

Bioengineering and Biotechnology. 2016;**2**(1):24-38

[41] Arun G, Eyini M, Gunasekaran P. Green synthesis of silver nanoparticles using the mushroom fungus schizophyllum commune and its biomedical applications. Biotechnology and Bioprocess Engineering. 2014;**19**:1083-1090

[42] Barapatre A, Aadil KR, Jha H. Synergistic antibacterial and antibiofilm activity of silver nanoparticles biosynthesized by lignin degrading fungus. Bioresources and Bioprocessing. 2016;**3**(8):1-13

[43] Birla SS, Tiwari VV, Gade AK, Ingle AP, Yadav AP, Rai MK. Fabrication of silver nanoparticles by phoma glomerata and its combined effect against *Escherichia coli*, *Pseudomonas aeruginosa* and *Staphylococcus aureus*. Letters in Applied Microbiology. 2009;**48**:173-179

[44] Subbaiya R, Selvam M. Synthesis and characterization of silver nanoparticles from streptomyces olivaceus sp-1392 and its anticancerous activity against non-small cell lung carcinoma cell line (NCI-H460). Current Nanoscience. 2014;**10**:243-249

[45] El-Sonbaty SM. Fungus-mediated synthesis of silver nanoparticles and evaluation of antitumor activity. Cancer Nano. 2013;**4**:73-79

[46] Gade AK, Bonde AP, Marcato PD, Duran N, Rai MK. Exploitation of *Aspergillus niger* for synthesis of silver nanoparticles. Biobased Materials and Bioenergy. 2008;**2**:243-247

[47] Govindappa M, Farheen H, Chandrappa CP, Channabasava R, Rai RV, Raghavendra VB. Mycosynthesis of silver nanoparticles using extract of endophytic fungi, penicillium species of glycosmis mauritiana and its antioxidant, antimicrobial, anti-inflammatory and tyrokinase inhibitory activity. Advances in Natural Sciences: Nanoscience and Nanotechnology. 2016;**7**:1-10

[48] Rajam KS, Rani ME, Gunaseeli R, Munavar MH. Extracellular synthesis of silver nanoparticles by the fungus *Emericella nidulans* EV4 and its application. Indian Journal of Experimental Biology. 2017;**55**:262-265

[49] Kulkarni RR, Shaiwale NS, Deobafkar DN, Deobagkar DD. Synthesis and extracellular accumulation of silver nanoparticles by employing radiation-resistant deinococcus radiodurans, their characterization, and determination of bioactivity. International Journal of Nanomedicine. 2015;**10**:963-974

[50] Netala VR, Bethu MS, Pushpalatha B, Baki VB, Aishwarya S, Rao JV, et al. Biogenesis of silver nanoparticles using endophytic fungus pestalotiopsis microspora and evaluation of their antioxidant and anticancer activities. International Journal of Nanomedicine. 2016;**11**:5683-5696

[51] Kaler A, Jain S, Banerjee UC. Green and rapid synthesis of anticancerous silver nanoparticles by *Saccharomyces boulardii* and insight into mechanism of nanoparticle synthesis. Biomedical Research International. 2013. DOI: 10.1155/2013/872940

[52] Durairaj B, Muthu S, Shanthi P. Larvicidal potential of fungi based silver nanoparticles against *Culex quinquefasciatus* larvae (ii and iii instar). Journal of Pharmacology and Toxicological Studies. 2014;**2**(4):42-49

[53] Ma L, Su W, Liu JX, Zeng XX, Huang Z, Li W, et al. Optimization for extracellular biosynthesis of silver nanoparticles by *Penicillium aculeatum* Su1 and their antimicrobial activity and cytotoxic effect compared with silver ions. Materials Science and Engineering:

C. 2017;**77**:963-971. DOI: 10.1016/j.msec.2017.03.294

[54] Manivasagan P, Venkatesan J, Senthilkumar K, Sivakumar K, Kim S-K. Biosynthesis, antimicrobial and cytotoxic effect of silver nanoparticles using a novel nocardiopsis sp. MBRC-1. Biomedical Research International. 2013. DOI: 10.1155/2013/287638

[55] Rahimi G, Alizadeh F, Khodavandi A. Mycosynthesis of silver nanoparticles from *Candida albicans* and its antibacterial activity against *Escherichia coli* and *Staphylococcus aureus*. Tropical Journal of Pharmaceutical Research. 2016;**15**(2):371-375

[56] Rajora N, Kaushik S, Jyoti A, Kothari SL. Rapid synthesis of silver nanoparticles by *Pseudomonas stutzeri* isolated from textile soil under optimised conditions and evaluation of their antimicrobial and cytotoxicity properties. IET Nanobiotechnology. 2016;**10**(6):367-373

[57] Shanmugasundaram T, Balagurunathan R. Mosquito larvicidal activity of silver nanoparticles synthesized using actinobacterium, Streptomyces sp. M25 against Anopheles subpictus, *Culex quinquefasciatus* and *Aedes aegypti*. Journal of Parasite Discovery. 2013;**39**:677-684. DOI: 10.1007/s12639-013-0412-4

[58] Kalaivani K, Prabhu P. Biosynthesis of silver nanoparticles using lactobacillus acidophilus and white rot fungus—A comparative study. International Journal of Advance Research, Ideas and Innovations in Technology. 2017;**3**(2):299-306

[59] Verma VC, Kharwar RN, Gange AC. Biosynthesis of antimicrobial silver nanoparticles by the endophytic fungus *Aspergillus clavatus*. Nanomedicine. 2010;**5**(1):33-40

[60] Waghmare SR, Mulla MN, Marathe SR, Sonawane KD. Ecofriendly production of silver nanoparticles using *Candida utilis* and its mechanistic action against pathogenic microorganisms. 3 Biotech. 2015;**5**:33-38. DOI: 10.1007/s13205-014-0196-y

[61] Alshehri AH, Jakubowska M, Mlozniak A, Horaczek M, Rudka D, Free C, et al. Enhanced electrical conductivity of silver nanoparticles for high frequency electronic applications. Applied Materials and Interfaces. 2012;**4**:7007-7010

[62] Chen D, Qiao X, Qiu X, Chen J. Synthesis of electrical properties of uniform silver nanoparticles for electronic applications. Journal of Material Science. 2009;**44**:1076-1081

[63] Yang X, He W, Wang S, Zhou G, Tang Y. Preparation and properties of a novel electrically conductive adhesive using a composite of silver nanorods, silver nanoparticles and modified epoxy resin. Journal of Material Science: Materials in Electronics. 2012;**23**:108-114

[64] Wu JT, Hsu SL. Preparation of triethylamine stabilized silver nanoparticles for low temperature sintering. Journal of Nanoparticle Research. 2011;**13**(1):3877-3883

[65] Ye L, Lai Z, Liu J, Tholen A. Effect of silver particle size on electrical conductivity of isotropically conductive adhesives. IEEE Transactions on Electronics Packaging Manufacturing. 1999;**22**(4):299-302

[66] Dankovich TA. Microwave assisted incorporation of silver nanoparticles in paper for point of use water purification. Environmental Science: Nano. 2014;**1**(4):367-378

[67] Park S, Ko YS, Jung H, Lee C, Woo K, Ko G. Disinfection of waterborne viruses using silver nanoparticle-decorated silica hybrid

composites in water environments. Science of the Total Environment. 2018;**625**:477-485

[68] Abu-Elala NM, Attia MM, Abd-Elsalam RM. Chitosan-silver nanocomposite in goldfish aquaria: New prospective for *Lernaea cyprinacea* control. Biomacromolecules. 2018;**111**:614-622. DOI: 10.1016/j.ijbiomac.2018.12.133

[69] Sanago R, Maity S, Mehta RK. Plasmonic effect due to silver nanoaparticles on silicon solar cell. Procedia Computer Science. 2016;**92**:549-553

[70] Li Z, Zhang Y, Ye J, Guo M, Chen J, Chem W. Nanozymatic glucose biosnensors based on silver nanoparticles deposited on TiO_2 Nanotubes. Journal of Nanotechnology. 2016. DOI: 10.1155/2016/9454830

[71] Ruth C, Loren WA, Mariaa GMP. Development of DNA biosensor based on silver nanoparticles UV-Vis absorption spectra for *Escherichia coli* detection. In: The 3rd International Conference on Biological Science. 2013. DOI: 10.18502/kls.v2i1.180

[72] Mahmudin L, Suharyadi E, Utomo ABS, Abraha K. Optical properties of silver nanoparticles for surface plasmon resonance (SPR)-based biosensor applications. Journal of Modern Physics. 2015;**6**:1071-1076

[73] Sistani P, Sofimaryo L, Masoudi ZR, Sayad A, Rahimzadeh R. A penicillin biosensor by using silver nanoparticles. 2014;**9**:6201-6212

[74] Tung NH, Chikae M, Ukita Y, Viet PH, Takamura Y. Sensing technique of silver nanoparticles as labels for immunoassay using liquid electrode plasma atomic emission spectrometry. Analytical Chemistry. 2014;**84**:1210-1213

[75] Duran N, Marcato PD, Souza GD, Alves OL, Esposito E. Antibacterial effect of silver nanoparticles produced by fungal process on textile fabrics and their effluent treatment. Journal of Biomedical Nanotechnology. 2007;**3**:203-208

[76] Hua X, Li H-W, Long Y-T. Investigation of silver nanoparticle induced lipids changes on single cell surface by ToF-SIMS. Analytical Chemistry. 2018;**90**:1072-1076. DOI: 10.1021/acs.analchem.7b04591

[77] Jung YJ, Metreveli G, Park C-B, Baik S, Schaumann GE. Implications of pony lake fulvic acid for the aggregation and dissolution of oppositely charged surface-coated silver nanoparticles and their ecotoxicological effects on *Daphnia magna*. Environmental Science and Technology. 2018;**52**:438-435. DOI: 10.1021/acs.est.7b04635

[78] Guilger M, Stigliani TP, Bilesky-jose N, Grillo R, Abhilash PC, Fraceto LF, et al. Biogenic silver nanoparticles based on trichoderma harzianum: synthesis, characterization, toxicity evaluation and biological activity. Scientific Reports. 2017. DOI: 10.1038/srep44421

[79] Babu MY, Devi VJ, Ramakritinan CM, Umarani R, Taredahali N, Kumaraguru AK Application of biosynthesized silver nanoparticles in agricultural and marine pest control. Current Nanoscience. 2014;**10**:1-9

[80] Chen JC, Lin ZH, Ma XX. Evidence of the production of silver nanoparticles via pretreatment of phoma sp.3.2883 with silver nitrate. Letters in Applied Microbiology. 2003;**37**:105-108

[81] Du L, Xu Q, Huang M, Xian L, Feng JX. Synthesis of small silver nanoparticles under light radiation by fungus penicillium oxalicum and its application for the catalytic reduction of methylene blue. Materials Chemistry and Physics. 2015;**160**:40-47

[82] Otari SV, Patil RM, Nadaf NH, Ghosh SJ, Pawar SH. Green synthesis of silver nanoparticles by microorganism using organic pollutant: its antimicrobial and catalytic application. Environmental Science Pollution Research. 2014;**21**:1503-1513. DOI: 10.1007/s11356-013-1764-0

[83] Zaheer Z. Biogenic synthesis, optical, catalytic, and in vitro antimicrobial potential of Ag-nanoparticles prepared using Palm date fruit extract. Journal of Phytochemistry and Photobiology B: Biology. 2018;**178**:584-592. DOI: 10.1016/j.jphotobiol.2017.12.002

[84] Soni N, Prakash S. Possible mosquito control by silver nanoparticles synthesized by soil fungus (Aspergillus niger 2587). Advances in Nanoparticles. 2013;**2**:125-132

[85] Salunkhe RB, Patil SV, Patil CD, Salunke BK. Larvicidal potential of silver nanoparticles synthesized using fungus cochliobolus lunatus against *Aedes aegypti* (Linnaeus, 1762) and anopheles stephensi liston (diptera; culicidae). Parasitology Research. 2011;**109**:823-831

[86] Suresh G, Gunasekar PH, Kokila D, Prabhu D, Dinesh D, Ravichandran N, et al. Green synthesis of silver nanoparticles using delphinium denudatum root extract exhibits antibacterial and mosquito larvicidal activities. Spectrochimica Acta Part A: Molecular and Biomolecular Spectroscopy. 2014;**127**:61-66

Chapter 9

Copper Complexes as Influenza Antivirals: Reduced Zebrafish Toxicity

Kelly L. McGuire, Jon Hogge, Aidan Hintze, Nathan Liddle, Nicole Nelson, Jordan Pollock, Austin Brown, Stephen Facer, Steven Walker, Johnny Lynch, Roger G. Harrison and David D. Busath

Abstract

Copper complexes have previously been developed to target His37 in influenza M2 and are effective blockers of both the wild type (WT) and the amantadine-resistant M2S31N. Here, we report that the complexes were much less toxic to zebrafish than $CuCl_2$. In addition, we characterized albumin binding, mutagenicity, and virus resistance formation of these metal complexes, and employed steered molecular dynamics simulations to explore whether the complexes would fit in M2. We also examined their anti-viral efficacy in a multi-generation cell culture assay to extend the previous work with an initial-infection assay, discovering that this is complicated by cell culture medium components. The number of copper ions binding to bovine serum albumin (BSA) correlates well with the number of surface histidines and BSA binding affinity is low compared to M2. No mutagenicity of the complexes was observed when compared to sodium azide. After 10 passages of virus in MDCK culture, the EC_{50} was unchanged for each of the complexes, i.e. resistance did not develop. The simulations revealed that the compounds fit well in the M2 channel, much like amantadine.

Keywords: medicinal metals, proton transport, plaque assay, Ames, CTX, CPE

1. Introduction

The influenza A M2 protein is a homotetrameric channel [1] that is particularly selective for protons [2] and is essential for uncoating of the virus [3]. The proton selectivity is due to the cluster of His37 imidazole side chains in the channel [4, 5]. This channel has been a primary antiviral target. Amantadine (AMT) and rimantadine (RMT) were highly successful as M2 blockers [6–8], but they became ineffective in 2005 when a mutation from serine to asparagine at residue 31 (S31N) in M2 occurred [9, 10].

Attempts have been made to develop variants of AMT, RMT and others that could block the V27A, L26F, or S31N mutations [11–14]. We explored a different approach that could, in theory, target all functional forms of M2 [15].

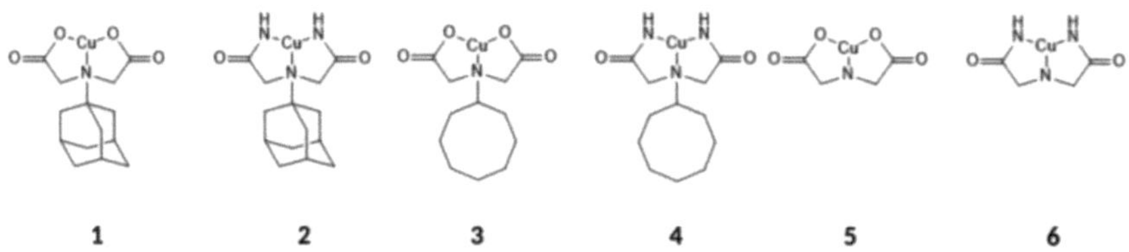

Figure 1.
Copper complexes with the functional groups iminodiacetate or iminodiacetamide extended via the amine to either AMT or CO.

Drawing from the observation that divalent cations, particularly copper, block M2 current [16] binding in the His37-Trp41 side chain quadruplex [17], divalent copper complexes of AMT were synthesized and found to be effective influenza A inhibitors with reduced cytotoxicity compared to $CuCl_2$ [15]. Because Cu^{2+} binds strongly to imidazole, it was suggested that the Cu^{2+} complexes also block M2 through His37-imidazole binding. In addition, the His37 cluster is highly conserved in nature [18], making it a prime target in the M2 channel.

The copper ligands developed were based on AMT and the lesser-known, equally effective M2 WT blocker, cyclooctylamine (CO) [19, 20], and extended via the amine with the functional groups iminodiacetate or iminodiacetamide. Six Cu^{2+} complexes (**Figure 1**) were synthesized and characterized using NMR, IR, MS, UV-Vis, and ICP-MS. The complexes demonstrated H37-specific block of M2 current in two electrode voltage-clamp (TEVC) studies with low μM potencies. The copper-free ligands did not show proton current block, demonstrating that the copper was key to the current-blocking process [15].

Because of the reduced toxicity to cultured cells found previously, we were interested to learn whether the six metal complexes were toxic to simple organisms. Zebrafish embryos were chosen because they have immune and nervous systems similar in many ways to more advanced organisms, because they are in an early, vulnerable stage of development, and because the compounds are readily administered at infection-relevant concentrations in their bathwater. We also explored and report additional properties of these copper complexes, including their efficacies in the cytopathic effect antiviral assay, their binding to albumin, mutagenicity testing in a bacterial assay, virus resistance development when passaged with cell culture in the presence of the compounds, and molecular dynamics simulations to explore how well the compounds fit in the M2 channel.

2. Materials and methods

2.1 Cytopathic effect assay

Confluent MDCK cells were transferred into 60 wells of a 96-well plate in DMEM (Gibco Thermo Scientific Waltham, MA, 4.5 g/L D-Glucose) with 5% Fetal Bovine Serum (FBS, Hyclone, Logan, UT). The cells were washed with a diluted solution of 50% SEM/50% serumless DMEM. SEM (simple electrolyte medium) consists of 4.33 g NaCl, 0.244 g KCl, 0.103 g $CaCl_2{\cdot}2H_2O$, $MgCl_2{\cdot}6H_2O$, $Na_2HPO_4{\cdot}7H_2O$, $NaH_2PO_4{\cdot}H_2O$ in 500 ml H_2O. The cells were incubated for an hour with activated A/WS/33 virus and then the media with virus was removed. The SEM/serumless DMEM with 100 μM metal complex was added to six wells. The complexes were then serial-diluted in two-fold increments six times. Six wells were used as positive controls with no complex or virus added. Six wells were used as negative controls with only virus added and no complex. About 80 μM ribavirin (Sigma-Aldrich, St. Louis, MO) was added to six wells as a positive control. The plates were incubated for 48 h at 33C.

The crystal violet staining technique described previously [21] was used to determine the fraction of cells that survived the exposure to the virus. After 48 h, the test medium was removed, and the cells were washed three times with 150 μl PBS. The cells were stained for 10 min with 50 μl crystal violet solution (0.03% crystal violet (w/v) in 20% methanol). The cells were then washed three times with 150 μl distilled water before adding 100 μl lysis buffer. After 20 min, the optical density (OD) of each well was measured at 590 or 620 nm and averaged over the set of six wells for each concentration.

Because viral dosing was sufficient to eliminate essentially all cells in treatment-free controls, their average OD was subtracted as baseline from the average of the treated well ODs. The result was divided by the average of the uninfected control well ODs to obtain a normalized vitality. Because the vitality can be affected by both reduction of virus cytopathic effect and increase of treatment toxicity as concentration is increased, we fitted the normalized concentration-dependent vitality, V(C), with a joint probability function:

$$V(C) = \frac{1}{1+\left(\frac{EC_{50}}{C}\right)^{n_1}} \frac{1}{1+\left(\frac{C}{CC_{50}}\right)^{n_2}} \quad (1)$$

Here, EC_{50} is the 50% effective dose of treatment that prevents viral cytopathic effect, CC_{50} is the 50% cytotoxic dose of the treatment, and n_1 and n_2 are their respective Hill coefficients. If the selectivity index, CC_{50}/EC_{50}, and the Hill coefficients are sufficiently high, this function rises to unity at doses that are sufficient to prevent viral replication but below toxic levels. Non-linear least squares fitting weighted with standard errors of means was done with the Marquardt algorithm in KaleidaGraph4 (Synergy Software, Reading, PA). In practice, it was necessary to fix the Hill coefficients to evaluate the effective doses, then manually adjust the Hill coefficients to improve the fit (due to low numbers of data points). Hence, the reported standard errors of the parameters obtained from the error matrix may be underestimated.

2.2 Protein binding assay

Each copper complex was dissolved in 25 ml of water to obtain a 1 mM and 800 μM solution. All water used in the protein binding assay was collected from a Millipore first-generation beige Milli-Q system. These solutions were sonicated until the crystals were fully dissolved. Four 1:2 serial dilutions were performed from the 800 μM solution to obtain 400, 200, 100, and 50 μM solutions, and a 1:5 dilution was performed from the 50 μM solution to obtain a 10 μM solution. 13.3 mg of BSA was then dissolved in 10 ml of each solution. The solutions were mixed thoroughly and allowed to stand at room temperature for approximately 20 min.

Spin filtration was performed using a swinging bucket rotor at 4000 rpm for 6 min. The spin filters used were Amicon Ultra-15 centrifugal filters. The filtrates from each spin were collected to test for copper content in ICP-MS. Solutions for ICP-MS were prepared from both the original solutions and the filtrates. For each solution, 1 ml of solution was added to 1 ml of 4% HNO3 and 8 ml of 2% HNO_3 to obtain a 1:10 dilution of each solution in 2% HNO_3. Nitric acid used for ICP-MS analysis was OmniTrace trace-metal grade obtained from EMD Millipore Corporation. We used BSA to model copper binding histidine in solution and calculate relative dissociation constants (K_d) for each complex. Copper concentrations were obtained using ICP-MS. The data was fit to Eq. **(2)** to estimate K_d and the number of binding sites, n.

$$\frac{[Cu]_t - [Cu]_f}{[P]_t} = \frac{n * [Cu]_f}{K_d + [Cu]_f} \tag{2}$$

2.3 Zebrafish toxicity test

Following an approved BYU IACUC protocol, two AB wild-type male and female zebrafish were placed in an embryo media filled tank. The fish remained in a light and temperature-controlled facility until the following morning. Later that day, the fish were transferred into original tank. Embryos were moved into embryo media filled petri dishes (60 embryos/dish) and housed in an incubator for 2 days. Media was changed daily.

Fish embryos were dechorionated at 48 hpf. In a multi-well plate, 10 embryos were selected and 5 were added to each of two wells for each concentration with fresh embryo media. Drug solution (0–200 μM) was then added to test toxicity and observed over 5 days. Drug solutions were changed daily. After 5 days, the fish were scored using a morbidity scale (**Table 1**) indicating response, spine shape, edema, equilibrium, and death. The average for each complex was normalized using the maximal morbidity score of 50/well. The fish were then euthanized.

2.4 Ames testing

The Modified Ames ISO kit (Environmental Bio-Detection Products Inc., Mississauga, ON) was used with *S. typhimurium* TA100 (no S9 fraction).

The complexes were compared against the mutagenicity of a positive control (NaN_3) and vehicle (water). The complexes were serially diluted 1:2 to compare the complexes' mutagenic ability at each of six concentrations.

TA100 was hydrated and incubated with histidine overnight at 37°C. Following the kit's instructions, in 96-well plates' exposure solution, diluted bacteria mix, and serial two-fold dilutions of complexes were combined with reversion media containing Bromocresol Purple, which serves as a pH indicator to identify infected wells. The 96 well plates were incubated for 6 days at 37°C without agitation. When a sample is mutagenic, it will revert the bacteria to WT, causing the media to turn slightly acidic and show a yellow color.

The number of reverted wells with complex was compared to the average number of reverted wells in the negative control. Significance was calculated using a one-tailed t-test.

2.5 Simulations

The 2KQT M2 structure was used and oriented in a DMPC lipid bilayer with a center-of-mass harmonic constraint. The copper complexes were oriented such that the copper was near (~2.0 Å) at least one of the four imidazole nitrogens. Water molecules within 2.2 Å of the complexes were deleted. The protein-bilayer system was solvated with a tetragonal 60 Å × 60 Å × 90 Å water box as shown in **Figure 2**. The system was minimized for 1000 steps of steepest descent and heated to 300 K. The M2 channel was equilibrated for 1 ns. The complexes were pulled using a constant force for 10 ps during the production runs. Frames were saved every 50 steps, which is every 50 fs, of production for a total of 200 frames. Standard CHARMM version 37b1 parameters were used. Copper dihedral parameters were created using a 20 kcal/mol/rad^2 energy penalty, which kept a conservatively rigid structure throughout the channel (**Table A1**).

Zebrafish scoring indicators				
Morbidity points	0	1	2	3
Equilibrium	Upright position	Lying on side	NA	NA
Response	Quick escape	Sluggish escape	No escape	NA
Spine shape	Straight	Slightly curved	Strongly curved	NA
Edema	None	1 place and minor	2 places or major	2 places and major
Death	=10			

Five fish per group in each of two wells.

Table 1.
Scoring indicators observed daily for 5 days.

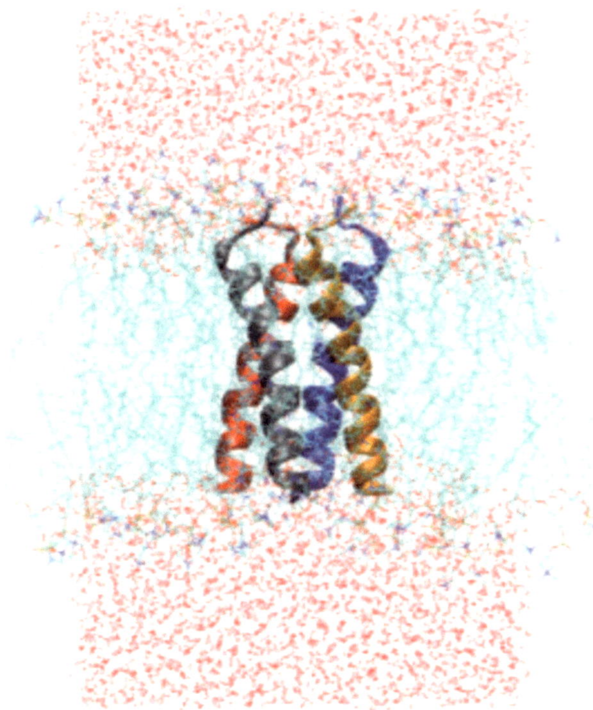

Figure 2.
Solvated DMPC-2KQT system.

The distance between imidazole nitrogens and copper on the complexes was calculated using CHARMM's CORREL subroutine for each frame. The time for each complex was recorded when the copper reached 30 Å away from the imidazole nitrogens. This distance was chosen to represent the complex leaving the mouth of the channel.

2.5.1 Decisions affecting pulling force

The pulling force for each of the complexes was determined by normalizing the pulling forces to a 2.34 nN pulling force on AMT. The 2.34 nN force allowed comparisons to be made between compounds as they left the channel on the 10 ps timescale.

These steered molecular dynamics (SMD) simulations were analyzed by computing the mean and standard deviation of five independent. The five independent simulations were assigned random starting velocities and then analyzed to explore the time needed to pass the 30 Å threshold relative to the starting point from the copper atom on the complex.

The analysis examined whether the pulling forces, copper ligation mechanism, or scaffold (CO, AMT, or neither), significantly affected the exit times relative to free Cu^{2+}.

2.6 Miniplaque assays, resistance testing, and sequencing

MDCK cells were seeded into a six-well plate and grown in Dulbecco's Modified Eagles Medium (DMEM, Sigma-Aldrich, St. Louis, MO) augmented by 5% with fetal bovine serum (FBS, Hyclone, Logan, UT) until confluent. After 48 h, the growth media was removed and replaced with DMEM. At this point the virus (A/CA/07/09) was introduced into the medium (200 pfu/ml) and allowed to adsorb for 1 h. The medium was then removed and replaced with fresh DMEM containing a specified concentration of complex and 5 ml of tosyl phenylalanyl chloromethyl ketone (TPCK)-treated trypsin (Thermo-Fischer Scientific, Waltham, MA, 1 mg/ml) was added to activate the virus. The plate was incubated at 33°C for 3 days. Then the medium was removed and centrifuged at 2000 rpm in order to remove cell debris. This virus-containing medium was then separated into 1-ml aliquots and frozen in Eppendorf tubes at −80°C. This process was repeated for each successive passage.

The concentration of virus was determined through an immunofluorescence assay (as previously described by [22]), which gave a multiplicity of infection (MOI) of 0.6. MDCK cells were seeded onto glass coverslips in vials containing 1 ml DMEM and trypsin in order to obtain 90% confluency after 24 h. The cells were allowed to grow overnight at 37°C, after which the growth medium was removed and replaced with DMEM. The sample of virus was then diluted by factors of 10, and the various dilutions of virus were stirred into the vials with coverslips. They grew at 33°C for 18 h. After this incubation period, the medium was removed, the cells were fixed with cold acetone (−80°C), and the coverslips were washed and stained with a fluorescein isothiocyanate labeled anti-IAV monoclonal antibody (Millipore Sigma, Burlington, MA, Cat. #5017). Excess antibody was washed off using a solution of 0.05% Tween20 in phosphate buffered saline and then again with distilled water. They were then viewed microscopically and individual infected cells (miniplaques) were counted.

This same process was followed in determining the new EC_{50} against the specific complex of each resistant strain. Except, 100 pfu of virus was used in each vial. Several different concentrations of the complex with which it was passaged were introduced into the vials, with concentrations ranging from 2 to 70 μM. The cells were infected with the virus in a solution of SEM rather than DMEM. The EC_{50} was calculated in KaleidaGraph using the Levenberg-Marquardt algorithm. The fitting parameters (sigmoidal function) were used to calculate the EC_{50} and the standard error of the mean.

To sequence the genome, the viral sample was concentrated 10-fold using a spin filter (VWR North America, Radnor, PA, Cat. #82031-352). After that, viral RNA was isolated using the QIAamp Viral RNA Mini Kit (Qiagen, Germantown, MD). The isolated RNA was stored at −20°C. RNA was then reverse-transcribed using Invitrogen's Superscript III One-Step RT-PCR Platinum Taq HiFi kit (Thermo-Fischer Scientific, Waltham, MA).

The resulting isolated DNA was stored at −20°C. The DNA was then amplified with PCR using the Phusion High-Fidelity PCR kit (New England Biolabs, Ipswich, MA). The solution was purified using Qiagen's QIAquick PCR

Purification Kit (Qiagen, Germantown, MD). It was sequenced using custom forward (TGTAAAACGACGGCCAGTACGAAAAGCAGGTAG) and reverse (CAGGAAACAGCTATGACCAGTAGAAACAAGGTAGT) primers for the segment of the new DNA that codes for the M2 protein.

3. Results and discussion

3.1 Cytopathic effect assay

Although **1–4** had good potency against initial infections in the immunofluorescence (miniplaque) assay [15], the copper complexes had no effect in the cytopathic effect (CPE) assay with MDCK cells when dissolved in serumless DMEM. However, when the serumless DMEM was diluted with SEM, **1** (**Figure 3**) and to a lesser extent **3** (data not shown) exhibited cell protection. Using a dual-sigmoidal function curve fit, **1** has an EC_{50} of 0.9 ± 0.08 μM and a CC_{50} of 5.8 ± 0.37 μM. The submaximal efficacy is due to high cytotoxicity. The selectivity index for **1** is 6.44, given by the ratio of the CC_{50} and EC_{50}. The low EC_{50} compares favorably to the EC_{50} in the miniplaque assay, 6.7 ± 1.2 μM. In contrast, **3** has an EC_{50} greater than 100 μM, whereas its potency in the miniplaque assay was EC_{50} = 0.7 ± 0.1 μM, and **2** and **4–6** showed no effect, indicating that other factors were involved. The fact that some efficacy is observed when the medium is diluted with amino-acid free SEM suggests that free amino acids in non-dilute DMEM interfere with the copper complex efficacy.

3.2 Protein binding assay

To illustrate the potential of the metal complexes to bind to proteins, binding to BSA was measured in which a protein solution was mixed with various concentrations of a $CuCl_2$ or copper complex solution. The copper content of the original sample

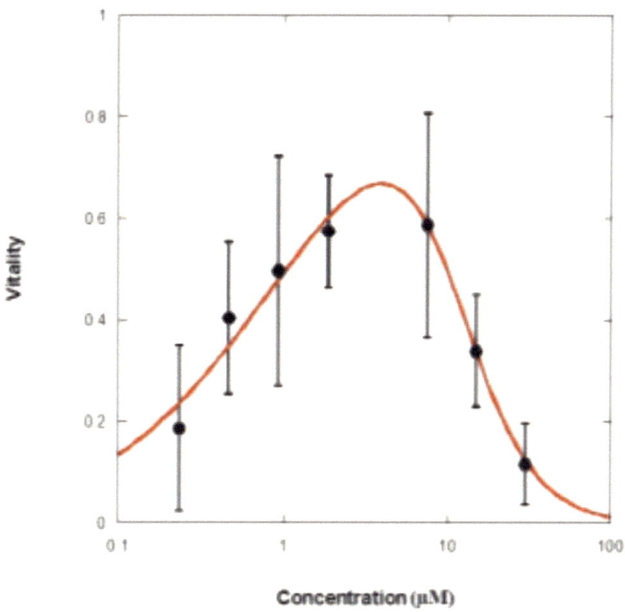

Figure 3.
CPE assay showing protective effect of 1 against A/WS/33 (M2 S31N) infection of MDCK cells using dilute medium (50% DMEM, 50% SEM). MOI 0.6; 48-h incubation.

was measured and compared to that of the filtrate. Taking the volume proportions into account, the "free copper concentration" in the filtrate relative to the "total copper concentration" in the original sample was fitted to a model assuming that each protein molecule had n equivalent copper or copper complex binding sites. **Table 2** shows the best fit K_d values, assuming that each albumin monomer has n equivalent binding sites. The two parameters interacted and were therefore poorly constrained in the optimization of the deviations squared, but **Table 2** indicates that the number of binding sites is well above 10, consistent with the count of 13 surface histidines in monomeric albumin (**Figure A1**). Complexes **1, 3,** and **5** have larger K_d values compared to that of $CuCl_2$ (59.1 μM). This indicates that the ligands on the metal complexes reduce the binding affinity for albumin binding sites, but also still allow for substantial binding. It is also consistent with the electrophysiology results for blocking through binding of copper complex to the His37 cluster in the M2 channel.

BSA has 13 surface histidines (**Figure A1**), however, all of the fits optimized n at >13 copper binding sites. This difference could suggest non-specific binding to other sites on BSA. The high K_d's for the complexes relative to $CuCl_2$ indicate that the complexes remain intact during binding to BSA. The binding the copper complexes to BSA is very weak compared to that of the $M_2S_{31}N$ (AMT resistant) channel, where block was ~80% for **1** and **3** after 57 and 27 min perfusion, respectively. This suggests that protein binding *in vivo* would be a minor concern. However, it is clear that binding by non-M2 proteins is detectable and, given their large quantity inside and outside the blood, they could limit access of the copper complexes to virus.

3.3 Zebrafish toxicity test

Toxicity was evaluated for zebrafish exposed to various concentrations of $CuCl_2$ or copper complex (**1–5**) added as methanolic solutions to the embryo bath medium starting 48 h post fertilization (Day 0) (**Figure 4**). At 200 μM copper complex on day 1, compounds **2, 4,** and **5** show minimum toxicity effects, **1** and **3** show moderate toxicity including slow response to stimulation, slightly curved spine, and minor edema, whereas $CuCl_2$ causes major edema, strongly curved spine, no response to stimulation, and death. By day 2 at 200 μM, the toxicities of **1, 2, 4,** and **5** have increased moderately but still only moderate spine curvature and minor edema, while **3** causes slow response to stimulation, strongly curved spine, and moderate to major edema. By days 3, 4, and 5 at 200 μM, all but **5** show low or no response, strongly curved spines, major edema, and some death. The MeOH vehicle controls showed statistically insignificant toxicity.

Compared to $CuCl_2$, the copper complexes show less toxicity, suggesting that the ligands are coordinating to the copper and helping to reduce its toxicity through day 2 of high dosage. All of the copper complexes produce some toxicity in the zebra fish for all experimental concentrations, but compound **5** does not increase in

Complex	K_d (μM)	Sites (n)
$CuCl_2$	59.1	15
1	128.7	19
3	179.5	20
5	380	20*

*K_d values for representative compounds 1, 3, and 5.*The value of n was not well-constrained and was therefore fixed during the curve fit.*

Table 2.
Binding to bovine serum albumin results.

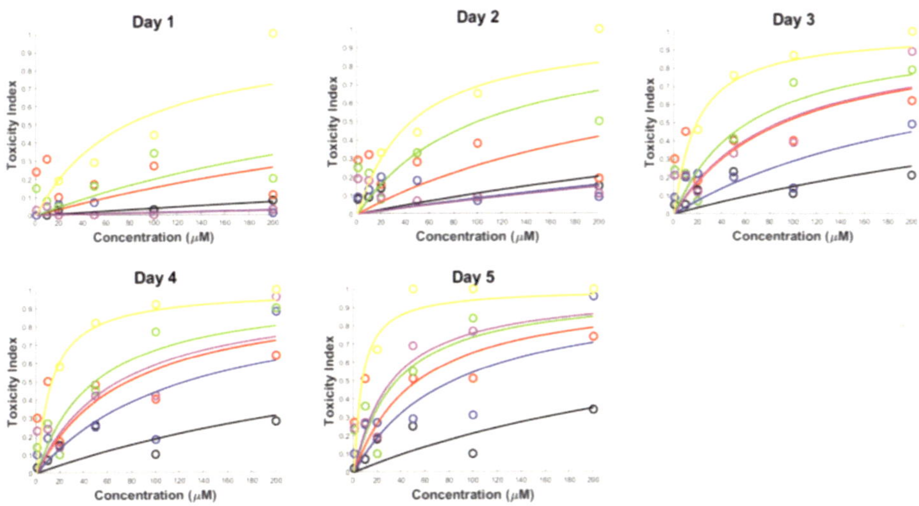

Figure 4.
Zebrafish toxicity of copper complexes. $CuCl_2$ (yellow), 1 (red), 2 (blue), 3 (green), 4 (magenta), 5 (black). (Arbitrary toxicity index, see Section 2).

toxicity over time as much as the other complexes. This suggests further testing and modification of compound **5** could lead to a safe anti-influenza A therapeutic.

3.4 Ames testing

The mutagenicity of the copper complexes was tested using the Ames test. **Table 3** shows the percent reversion out of 48 wells of three complexes. They were tested for mutagenicity against *S. Typhimurium* TA100, which strain of bacteria allows a test for mutagenicity caused by base-pair substitution and oxidative damage. The percent of revertant wells (reversion rate) was compared against the negative control and found to be statistically insignificant ($p > 0.01$). The positive control (NaN_3) had an average 91.7% reversion rate compared to the negative control's average rate of 43.8% ($p < 0.0001$). Complexes **1, 3,** and **4** did not show significant rate of reversion at any tested concentration compared to NaN_3. The copper complexes showed approximately the same reversion rates as the negative control after 6 days. Therefore, they do not cause mutagenicity due to base-pair substitution or oxidative damage.

Concentration (μM)	Complex		
	1	3	4
500	42	42	50
250	42	42	30
125	31	52	33
62.5	31	41	38
31.25	52	58	67
15.63	56	52	56
0	35	46	50

Percent of reversion out of 48 wells for compounds 1, 3, and 4 for concentrations between 15 and 500 μM. 0 is water with no complex.

Table 3.
Ames mutagenicity assay results.

3.5 Resistance testing and sequence

Because the putative target for the metal complexes, the His37 quadruplex, is highly conserved in nature and functionally critical for vRNP uncoating, we explored the propensity for virus resistance formation with passaging in MDCK cell cultures. Because the incubation had to be done in DMEM, which is known to inhibit complex efficacy, we used higher concentrations of complexes for the incubations such that the efficacy of block was projected to be ~50%, thus creating a concentration where mutation could occur. Ten passages (~5 weeks) of incubated virus in DMEM dosed with increasing metal complex concentrations (ranging from 50 to 100 μM) was chosen as a rigorous test. Resistance would be identifiable by an increase in miniplaque EC_{50} after passaging relative to the original value. As shown in **Table 4**, the new EC_{50} (column 3) is comparable to the original EC_{50} (column 2). Because none of the copper complexes significantly increased the EC_{50} after 5 weeks of incubation, we conclude that resistance is slow to develop. This contrasts with rapid resistance development when passaging with AMT [15].

The vRNA M segment was extracted from the passaged virus exposed to **3**, sequenced and compared to A/CA/07/2009 using a reverse-BLAST mechanism. The only base mutation discovered was G749A, which translates to the amino acid mutation G16E. This amino acid is positioned in the region of the channel entry that is outside of the membrane and unlikely to influence channel permeation. According to the results in the above table, this mutation did not confer resistance to this compound. We consider the occasionally observed natural M2 mutant G34E to be likely to escape block by these complexes. Although we did not see resistance develop in our assays, a more direct assessment of the G34E site mutation using electrophysiology might be instructive about resistance potential for these compounds in future studies.

3.6 MD simulations

Constant force steered molecular dynamics (MD) simulations were carried out to explore the steric limitations on metal complex exit from the M2 transmembrane domain AMT binding site. A 2.34 nN force was used to pull the complexes pass the 30 Å threshold and beyond the Val27 cluster within 10 ps. The 2.34 nN force gave a sufficient spread in leaving times to allow assessment of the ease of unbinding relative to AMT. For these simulations, the force was applied to the center-of-mass of the complex. Example trajectories for AMT (green) and **4** (yellow) are shown in **Figure 5**. The starting configurations (left) had the adamantyl groups of AMT and **4** superimposed with the copper atom of **4** oriented down, close to H37. This binding configuration was used for all of the metal complexes. V27 and H37 are shown as reference points along the channel.

Complex	Original A/CA/09 (μM)	10 passages with complex (μM)
1	6.9 ± 1.2	3.7 ± 0.5
2	4.9 ± 0.8	2.1 ± 1.1
3	0.7 ± 0.1	1.1 ± 0.4
4	11.6 ± 1.1	3.9 ± 6.8
5	8.2 ± 2.0	1.3 ± 0.2
6	4.4 ± 0.6	2.9 ± 0.3

DMEM was used for passage incubations and SEM for the miniplaque assays.

Table 4.
Miniplaque EC_{50} ± SE (EC_{50}) before (original) and after 10 passages of virus in MDCK cells.

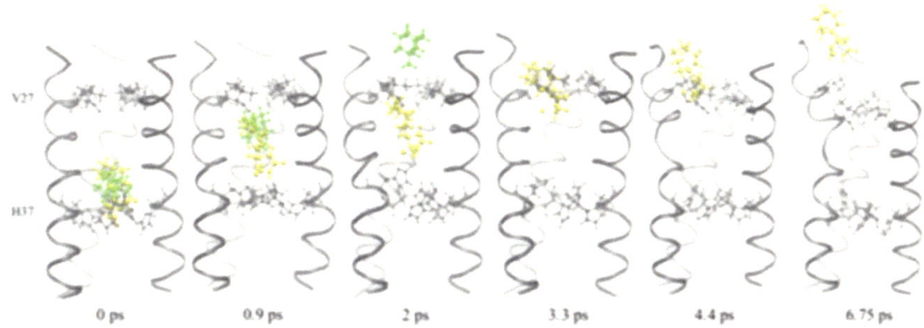

Figure 5.
Exit trajectories of AMT and 4 leaving M2 channel. Three of the four M2 TMD monomers from the simulation of 4 are displayed for reference. Valine 27 and Histidine 37 side chains are shown in grey, 4 is yellow and AMT is green. Both complexes are experiencing the same 2.34 nN pulling force.

Complex	Average time to leave (ps)
1	4.60 ± 1.14
2	6.85 ± 1.01
3	4.67 ± 1.07
4	7.05 ± 0.53
5	3.75 ± 0.89
6	4.00 ± 1.06
AMT	2.77 ± 0.26

Table 5.
Average time to leave (±standard deviation) the channel with a 2.34 nN force applied to the compound's center-of-mass.

Table 5 shows the average time to leave from five independent simulations (identical starting configurations, but randomly assigned atomic velocities) for each complex to pass the 30 Å threshold. All metal complexes took longer to leave the channel than AMT. AMT exited the channel in 2.77 ps. Complex **4** interacts with the V27 side chain and was the slowest compound to leave the channel, with its leaving time at 7.05 ps. By 4.4 and 6.75 ps, some distortion is seen on the protein subunit as **4** is pulled further out of the channel.

4. Conclusion

The copper complexes are relatively non-toxic in zebrafish embryos compared to $CuCl_2$ over a 5-day period. Also, they are efficacious in a 3-day assay (but with limitations due to serum protein binding and amino acid interference), are non-mutagenic compared to sodium azide, are slower to leave the M2 binding site compared to AMT, and, also compared to AMT, are not prone to resistance development. *In vivo* they would face competition with binding to other proteins and the therapeutic window is small. However, complexation of copper could be pharmacologically beneficial.

Further testing of these copper complexes should include isothermal titration calorimetry (ITC) experiments with influenza A M2 channel to obtain binding energies, two-electrode voltage clamp (TEVC) experiments to obtain rate constants of binding to M2, and testing in an animal model that more accurately represents the effects of the copper complexes on humans.

Appendix

See **Figure A1** and **Table A1**.

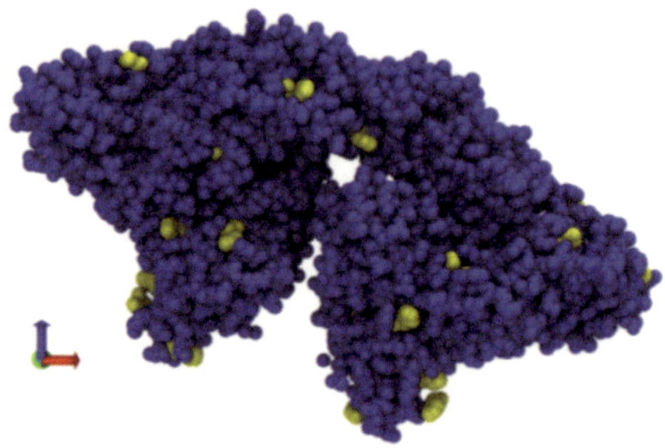

Figure A1.
Bovine serum albumin view (RCSB 3V03). Histidine side chains are colored yellow.

CU parameters			
Bond	**Kb (kcal/mol/Å2)**	**B0 (Å)**	
CU-N	270.2	2.026	
Angle	**Kθ (kcal/mole/rad2)**	**θ0 (degrees)**	
CT1-NPH-CU	96.150	128.05	
HB1-NPH-CU	30	123	
CT2-NH_3-CU	96.150	128.05	
H-NPH-CU	0	180	
NPH-CU-NH_3	14.39	90	
NPH-CU-NPH	14.39	90	
Dihedral	**Φ (kcal/mole/rad2)**	**Multiplicity**	**Delta (degrees)**
NPH-CU-NH_3-CT2	20	1	92.30
CU-NPH-C2-OB	20	2	169.1
CU-NH_3-CT1-CT2	20	1	−58.6
H-NPH-CU-NH_3	20	2	157.9
NPH-CU-NPH-H	20	2	161.2
H-NPH-CU-NH_3	20	2	157.9
C2-NPH-CU-NPH	20	2	18.8
CU-NPH-C2-CT2	20	2	8.0
CU-NH_3-CT2-HA	20	3	90.20
C2-CT2-NH_3-CU	20	2	−30.80
NPH-CU-NH_3-H	20	2	161.2

Table A1.
CHARMM copper parameters.

Author details

Kelly L. McGuire[1], Jon Hogge[1], Aidan Hintze[1], Nathan Liddle[1], Nicole Nelson[1], Jordan Pollock[1], Austin Brown[1], Stephen Facer[1], Steven Walker[1], Johnny Lynch[2], Roger G. Harrison[2*] and David D. Busath[1*]

1 Department of Physiology and Developmental Biology, Brigham Young University, Provo, UT, USA

2 Department of Chemistry and Biochemistry, Brigham Young University, Provo, UT, USA

*Address all correspondence to: roger_harrison@byu.edu and david_busath@byu.edu

References

[1] Busath DD. Influenza A M2: Channel or Transporter? Advances in Planar Lipid Bilayers and Liposomes. Vol. 10. Burlington: Academic Press; 2009. pp. 161-201

[2] Chizhmakov IV, Geraghty FM, Ogden DC, Hayhurst A, Antoniou M, Hay AJ. Selective proton permeability and pH regulation of the influenza virus M2 channel expressed in mouse erythroleukaemia cells. The Journal of Physiology. 1996;**494**(Pt 2):329-336

[3] Helenius A. Unpacking the incoming influenza virus. Cell. 1992;**69**(4):577-578

[4] Venkataraman P, Lamb RA, Pinto LH. Chemical rescue of histidine selectivity filter mutants of the M2 ion channel of influenza A virus. The Journal of Biological Chemistry. 2005;**280**(22):21463-21472

[5] Wang C, Lamb RA, Pinto LH. Activation of the M2 ion channel of influenza virus: A role for the transmembrane domain histidine residue. Biophysical Journal. 1995;**69**(4):1363-1371

[6] Davies WL, Grunert RR, Haff RF, McGahen JW, Neumayer EM, Paulshock M, et al. Antiviral activity of 1-adamantanamine (amantadine). Science. 1964;**144**(3620):862-863

[7] Krylov VF, Alekseeva AA, Liarskaia T, Poliakova TG, Kupriashina LM. Therapeutic effectiveness of bonafton and rimantadine in influenza. Voprosy Virusologii. 1976;(2):186-191

[8] Wang C, Takeuchi K, Pinto LH, Lamb RA. Ion channel activity of influenza A virus M2 protein: Characterization of the amantadine block. Journal of Virology. 1993;**67**(9):5585-5594

[9] Hata M, Tsuzuki M, Goto Y, Kumagai N, Harada M, Hashimoto M, et al. High frequency of amantadine-resistant influenza A (H_3N_2) viruses in the 2005-2006 season and rapid detection of amantadine-resistant influenza A (H_3N_2) viruses by MAMA-PCR. Japanese Journal of Infectious Diseases. 2007;**60**(4):202-204

[10] Krumbholz A, Schmidtke M, Bergmann S, Motzke S, Bauer K, Stech J, et al. High prevalence of amantadine resistance among circulating European porcine influenza A viruses. The Journal of General Virology. 2009;**90**(Pt 4): 900-908

[11] Balannik V, Wang J, Ohigashi Y, Jing X, Magavern E, Lamb RA, et al. Design and pharmacological characterization of inhibitors of amantadine-resistant mutants of the M2 ion channel of influenza A virus. Biochemistry. 2009;**48**(50):11872-11882

[12] Wang J, Ma C, Wang J, Jo H, Canturk B, Fiorin G, et al. Discovery of novel dual inhibitors of the wild-type and the most prevalent drug-resistant mutant, $S_{31}N$, of the M2 proton channel from influenza A virus. Journal of Medicinal Chemistry. 2013;**56**(7):2804-2812

[13] Wu Y, Canturk B, Jo H, Ma C, Gianti E, Klein ML, et al. Flipping in the pore: Discovery of dual inhibitors that bind in different orientations to the wild-type versus the amantadine-resistant S31N mutant of the influenza A virus M2 proton channel. Journal of the American Chemical Society. 2014;**136**(52):17987-17995

[14] Zhao X, Jie Y, Rosenberg MR, Wan J, Zeng S, Cui W, et al. Design and synthesis of pinanamine derivatives as anti-influenza A M2 ion channel inhibitors. Antiviral Research. 2012;**96**(2):91-99

[15] Gordon NA, McGuire KL, Wallentine SK, Mohl GA, Lynch JD,

Harrison RG, et al. Divalent copper complexes as influenza A M2 inhibitors. Antiviral Research. 2017;**147**:100-106

[16] Gandhi CS, Shuck K, Lear JD, Dieckmann GR, DeGrado WF, Lamb RA, et al. Cu(II) inhibition of the proton translocation machinery of the influenza A virus M2 protein. The Journal of Biological Chemistry. 1999;**274**(9):5474-5482

[17] Su Y, Hu F, Hong M. Paramagnetic Cu(II) for probing membrane protein structure and function: Inhibition mechanism of the influenza M2 proton channel. Journal of the American Chemical Society. 2012;**134**(20):8693-8702

[18] Durrant MG, Eggett DL, Busath DD. Investigation of a recent rise of dual amantadine-resistance mutations in the influenza A M2 sequence. BMC Genetics. 2015;**16**(Suppl. 2):S3

[19] Pinto CA, Haff RF. Antival activity of cyclooctylamine hydrochloride in influenza virus-infected ferrets. Antimicrobial Agents and Chemotherapy. 1968;**8**:201-206

[20] Lin TI, Heider H, Schroeder C. Different modes of inhibition by adamantane amine derivatives and natural polyamines of the functionally reconstituted influenza virus M2 proton channel protein. The Journal of General Virology. 1997; **78**(Pt 4):767-774

[21] Schmidtke M, Schnittler U, Jahn B, Dahse HM, Stelzner A. A rapid assay for evaluation of antiviral activity against coxsackievirus B3, influenza virus A, and herpes simplex virus type 1. Journal of Virological Methods. 2001;**95**:133-143

[22] Kolocouris A, et al. Amino-adamantanes with Persistent in Vitro Efficacy against H1N1 (2009) Influenza A. Journal of Medicinal Chemistry. 2014;**57**(11):4629-4639